SpringerBriefs in Mathematics

SpringerBriefs in Mathematics showcases expositions in all areas of mathematics and applied mathematics. Manuscripts presenting new results or a single new result in a classical field, new field, or an emerging topic, applications, or bridges between new results and already published works, are encouraged. The series is intended for mathematicians and applied mathematicians.

More information about this series at http://www.springer.com/series/10030

Guangchen Wang • Zhen Wu • Jie Xiong

An Introduction to Optimal Control of FBSDE with Incomplete Information

Springer

Guangchen Wang
School of Control Science and Engineering
Shandong University
Jinan, China

Zhen Wu
School of Mathematics
Shandong University
Jinan, China

Jie Xiong
Department of Mathematics
Southern University of Science
and Technology
Shenzhen, China

ISSN 2191-8198 ISSN 2191-8201 (electronic)
SpringerBriefs in Mathematics
ISBN 978-3-319-79038-1 ISBN 978-3-319-79039-8 (eBook)
https://doi.org/10.1007/978-3-319-79039-8

Library of Congress Control Number: 2018937963

Mathematics Subject Classification: 49N10, 60H10, 60G35, 91G80, 93E11, 93E20

Printed on acid-free paper

This Springer imprint is published by the registered company Springer International Publishing AG part of Springer Nature.
The registered company address is: Gewerbestrasse 11, 6330 Cham, Switzerland

To our parents:

Yuanzhen Wang and Changxiang Zhang

Datong Wu and Jin Chen

Desui Xiong and Xiangyun Lu

Preface

Forward-backward stochastic differential equations (FBSDEs) are a special type of stochastic differential equations that have been studied extensively since the early 1990s (see, e.g., El Karoui et al. [19]). It provides a viable tool to solve certain stochastic control problems. Its applications have been found in many applied sciences, especially in quantitative finance as they can be used to describe effectively the dynamic behavior of the prices of the financial derivatives, investment portfolios, and other time-varying financial instruments that are stochastic in nature. It is also worth mentioning that one of the original motivations of a backward stochastic differential equation (BSDE), initiated by Bismut's early work in 1973 [10], is to describe the "adjoint equation" in the stochastic maximum principle.

The study of the optimal control problems for FBSDEs is of theoretical and practical significance. A standard assumption in the literature is that the stochastic noises in the model are observed completely. However, this is rarely the case in real-world situations. The optimal control problems under complete information are studied extensively. Nevertheless, very little is known about these problems when the information is not complete. The goal of this book is to fill this gap in the literature by presenting a systematical introduction and to attract more research attention to this exciting subject.

The main difficulty of the optimal control under partial observation is the circular dependency of the control process and the information filtration obtained from the observation process. Namely, the control process must be adapted to the observation filtration, while on the other hand the observation process depends on the control process. We will introduce a backward separation approach to decouple the filtering from the control problem.

The backward separation approach is applicable to many classes of nonlinear control systems such as the controlled FBSDEs. Combining the approach with a duality technique, we establish stochastic maximum principle (SMP) for optimality of a stochastic control for FBSDE, which is the conditional expectation of a certain Hamiltonian function with respect to the available information. In order to determine an optimal control, we further study some nonlinear filters of BSDEs and/or FBSDEs, which are different from the traditional filtering theory.

New control models arising from optimal premium, risk management, and recursive utility are provided. By the theoretical results introduced here, optimal feedback solutions of these control models are obtained.

This book is written in a style suitable for graduate students in mathematics and engineering with basic knowledge of stochastic process, optimal control, and mathematical finance. This book is organized as follows. In Chapter 1, we present a few motivating examples, formulate the general problem and various models, and provide an overview of this book. In Chapter 2, we consider the filtering problem for the BSDEs and the FBSDEs. In Chapter 3, we study the stochastic maximum principle for a general optimal control problem when the information filtration is given by a general family of sub-σ-fields. Chapter 4 is then devoted to the special case when the information filtration is provided by an observation process which itself depends on the control process. Finally, we further specialize the setting to linear state with quadratic lost functional (LQ problem) and apply to a concrete optimal premium problem motivated from actuary science. We introduce the BSDEs and the FBSDEs in the appendix for the convenience of the reader.

Wang acknowledges the financial support from the National Natural Science Foundation for Excellent Young Scholars of China (61422305), the National Natural Science Foundation of China (11371228), and the Natural Science Foundation for Distinguished Young Scholars of Shandong Province of China (JQ201418). Wu acknowledges the financial support from the National Natural Science Foundation of China (61573217), the National Natural Science Fund for Distinguished Young Scholars of China (11125102), the National High-level Personnel of Special Support Program, and the Changjiang Scholar Program of Chinese Education Ministry. Xiong acknowledges the financial support from FDCT 025/2016/A1 and start-up fund of Southern University of Science and Technology Y01286220. We would like to thank Hancheng Guo, Qizhu Liang, Shuaiqi Zhang, and Jiayu Zheng for their careful reading of the manuscript and for making many constructive suggestions and catching many typos from early versions of this manuscript. Finally, we would also like to thank three anonymous reviewers for their constructive suggestions which improve this book substantially.

Jinan, China Guangchen Wang
Jinan, China Zhen Wu
Shenzhen, China Jie Xiong
May 2017

Contents

Acronyms

BSDE	backward stochastic differential equations
BDG inequality	Burkholder–Davis–Gundy inequality
FBSDE	forward-backward stochastic differential equation
ODE	ordinary differential equation
SDE	stochastic differential equation
SMP	stochastic maximum principle
SPDE	stochastic partial differential equation

Chapter 1
Introduction

Stochastic optimal control with incomplete information is composed of filtering and control. The filtering part is related to two stochastic processes: signal and observation. The signal process is what we want to estimate based on the observation which provides the information we can use. Kalman–Bucy filtering is the most successful result in linear filtering theory, which was obtained by Kalman and Bucy [38]. Nonlinear filtering is much more difficult to study. There have been two essentially different approaches so far. One is based on the innovation process, an observable Brownian motion, with the martingale representation theorem. This theory achieved its culmination with the celebrated paper of Fujisaki et al. [25]. See also Liptser and Shiryayev [49] and Kallianpur [36] for a systematic account of this approach. Another approach was introduced by Duncan [18], Mortensen [56], and Zakai [112] independently, who derived a linear stochastic partial differential equation (SPDE) satisfied by the unnormalized conditional density function of the signal. This SPDE is called the Duncan–Mortensen–Zakai equation, or, simply, Zakai's equation. Unlike the Kalman–Bucy filtering, nonlinear filtering results in infinite-dimensional stochastic processes, whose analytical solutions are rarely available in general. Much effort has been devoted to finding finite-dimensional filters and numerical schemes. See, e.g., Beneš [5], Wonham [98], Xiong [104], and Bain and Crisan [2] for the development of this aspect.

Stochastic optimal control under complete information has been studied extensively. We refer the reader to Pham [69], Touzi [80], Yong and Zhou [109], and the references therein for more details toward maximum principle, dynamic programming, LQ control, and their applications to mathematical finance.

Stochastic optimal control with incomplete information is substantially more difficult comparing to its complete information counterpart. In the case of Gaussian system, partially observable optimal control can be addressed by the traditional separation principle. This principle allows one to first compute the filtering of the state process, and then to solve a complete information optimal control problem driven by the filtering. For non-Gaussian system, however, this principle is usually invalid,

G. Wang et al., *An Introduction to Optimal Control of FBSDE with Incomplete Information*, SpringerBriefs in Mathematics,
https://doi.org/10.1007/978-3-319-79039-8_1

mainly due to the fact that mean square error of filtering depends on control. In principle, we may turn this problem into a complete information optimal control driven by the Zakai's equation of the non-Gaussian state. This approach nevertheless leads to an optimal control problem in infinite dimensional space, which is difficult to solve. See, e.g., the monographs of Bensoussan [6] and Shen et al. [72] for systematic accounts. Therefore, new approaches and techniques need to be developed to study this problem.

Motivated by the fact mentioned above, we propose a backward separation approach in this book, which decouples control and filtering by first deducing optimal control and then computing optimal filtering. Since we use the original, finite dimensional state and observation equations to calculate the variation of the control problem, rather than Zakai's equation of the state which is infinite dimensional in general, lots of complicated stochastic calculus in infinite dimensional spaces are avoided, in contrast with Bensoussan [6]. Thus, the backward separation approach can be used to study the control problem in a more effective way. This is the *first main contribution* of this book. The backward separation approach is also applicable to other classes of nonlinear control systems such as the controlled forward backward stochastic differential equations (FBSDEs). Combining the approach with a duality technique, we establish a maximum principle (of Pontryagin's type) for the optimality of a stochastic control for the FBSDEs, which is the conditional expectation of a certain Hamiltonian function with respect to the available information. In order to determine an optimal control, we further study some nonlinear filters of backward stochastic differential equations (BSDEs) and/or FBSDEs, which are different from the traditional filtering theory. This can be regarded as the *second contribution* of this book. The *third contribution* is as follows. New control models arising from optimal premium, risk management, and recursive utility are provided. By the theoretical results introduced here, optimal feedback solutions of these control models are obtained.

The rest of this chapter is organized as follows. We first give a few motivating examples for stochastic optimal control with incomplete information. Then, we introduce several main models that arise in practical applications. Finally, we give an overview of the topics to be covered in this book.

1.1 Motivating Examples

In this section, we introduce three examples arising from different fields of application. The first example comes from wireless communication that was in fact the main motivation for filtering theory at its early stage. The second example comes from actuarial science where the cash-balance process affecting the stock price is not observed completely, instead, only the stock price itself is observed. The selection of the optimal premium rate must be based on the available information from the stock prices. Finally, in the third example, we introduce a risk minimizing problem from mathematical finance, where the economic quality is only partially observable through the stock prices. Although both examples are based on the available stock price information, the third example is different from the second that not only the large deviation from the benchmark of the control strategy is prevented, but also the risk of the economic quantity is minimized.

Throughout this book, we let $T > 0$ be a fixed time horizon, $\mathscr{F}_t$ be a natural filtration generated by a certain standard Brownian motion up to time t, and $\mathscr{F} = \mathscr{F}_T$. Denote by $(\Omega, \mathscr{F}, \mathbb{F} \equiv (\mathscr{F}_t)_{0\le t\le T}, \mathbb{P})$ a complete filtered probability space satisfying the usual conditions. Denote by $\mathbb{R}^m$ the m-dimensional Euclidean space, by $|\cdot|$ (resp. $\langle\cdot,\cdot\rangle$) the norm (resp. inner product) in a Euclidean space, by $A^\top$ the transpose of A, by f_x the partial derivative of f with respect to a variable x. Denote by $\mathscr{L}^2_{\mathbb{F}}(0,T;S)$ the set of all S-valued, $\mathscr{F}_t$-adapted, and square-integrable processes, and by $\mathscr{L}^2_{\mathscr{F}_T}(\Omega;S)$ the set of all S-valued, $\mathscr{F}_T$-measurable, and square-integrable random variables. Sometimes, the set S in $\mathscr{L}^2_{\mathbb{F}}(0,T;S)$ and $\mathscr{L}^2_{\mathscr{F}_T}(\Omega;S)$ are omitted for simplicity. Similar notations are used for other spaces and integrals.

1.1.1 Wireless Communication

A signal process $x(t)$ taking values in a space S is to be transmitted to a receiver. Because of the inaccuracy of measurements and the noisy environment, this signal is not directly observable. Instead, we observe a function $h(x(t))$ of the signal plus an m-dimensional white noise $n(t)$, i.e.,

$$y(t) = h(x(t)) + n(t),$$

where $h : S \to \mathbb{R}^m$, and $y(t)$ is called the observation process.

Note that as a function of the time variable t, the white noise exists only in the sense of generalized functions, it is the derivative of a Brownian motion that exists in the ordinary sense. To obtain a tractable mathematical model for the observation data, it is natural to consider the accumulated observation process

$$Y(t) = \int_0^t y(s)ds,$$

and thus, the observation model becomes a stochastic differential equation (SDE)

$$dY(t) = h(x(t))dt + dW(t), \quad Y(0) = 0,$$

where $W(\cdot)$ is an m-dimensional Brownian motion defined on $(\Omega, \mathscr{F}, (\mathscr{F}_t)_{0\le t\le T})$.

Note that no observable information is lost or gained by regarding $Y(\cdot)$ as our observation data instead of $y(\cdot)$. Define

$$\mathscr{F}^Y_t = \sigma\{Y(s); 0 \le s \le t\}.$$

Then the *filtering problem* is to estimate the signal $x(t)$ based on $\mathscr{F}^Y_t$ in an optimal way.

1.1.2 Optimal Premium Problem

Consider an insurance firm whose premium rate is denoted by $v(\cdot)$. An insurance portfolio consists of a large number of independent individual claims, none of which can affect the total returns significantly. Therefore, by the law of large numbers and the central limit theorem, the liability process $B^v(\cdot)$ can be approximated by (see, e.g., Norberg [59] for more details)

$$-dB^v(t) = (v(t) - b(t))dt + \sigma(t)dW(t).$$

Here $b(\cdot) > 0$ represents the expected liability per unit time due to premium loading; the premium rate $v(\cdot)$ acts as the control variable while the volatility rate $\sigma(\cdot) > 0$ measures the liability risk; $(W(\cdot), \tilde{W}(\cdot))$ is a 2-dimensional standard Brownian motion defined on a stochastic basis $(\Omega, \mathscr{F}, (\mathscr{F}_t)_{0\le t\le T}, \mathbb{P})$ with a correlation function $\rho(t)$; and $\mathscr{F}_t$ is the natural filtration generated by $(W(t), \tilde{W}(t))$. Note that we allow for $v(\cdot) < 0$ and this can be explained as the "reward rate" or "dividend rate" to the claim holder. Similar to Norberg [59], we assume that the insurance firm is not allowed to invest in the risky asset due to the supervisory regulations. Accordingly, the insurance firm only invests in a money account with compounded interest rate $\delta(\cdot)$, and hence, its cash-balance process $x^v(\cdot)$ is

$$x^v(t) = e^{\Delta(t)}\left(x_0 - \int_0^t e^{-\Delta(s)}dB^v(s)\right), \qquad x^v(0) = x_0,$$

where

$$\Delta(t) = \int_0^t \delta(s)ds,$$

and $x_0 \ge 0$ represents the initial reserve. According to Itô's formula,

$$\begin{cases} dx^v(t) = (\delta(t)x^v(t) + v(t) - b(t))dt + \sigma(t)dW(t), \\ x^v(0) = x_0. \end{cases} \tag{1.1}$$

Due to the physical inaccessibility to underlying economic parameters, discreteness of account information, or possible delay in the actual payments, it is possible that the cash-balance process is only observed partially by the linear factor model

$$\begin{cases} \frac{dS^v(t)}{S^v(t)} = (a + cx^v(t))dt + \gamma(t)d\tilde{W}(t), \\ S^v(0) = s_0. \end{cases} \tag{1.2}$$

Here both $c \neq 0$ and a are constants; and $\gamma(\cdot)$ and $\gamma^{-1}(\cdot)$ are both bounded, deterministic functions. Model (1.2) is such a special incomplete information model that it is of both application potentials and theoretical interests, as well as considerable analytical tractability. In fact, the model has been extensively studied in mathematical economics, see, e.g., Nagai and Peng [57] in risk sensitive control; Lakner [43] in portfolio optimization. However, the factor process is free of control in these works, and hence, essentially different from ours. One typical example of $S^v(\cdot)$ in practice is the stock price of the insurance firm. This is supported by Boswijk et al. [12], where the stock price is closely related to the underlying cash-balance process through the price-to-cash ratio which is linear. By (1.2), some incomplete information control problems arise naturally if we regard $x^v(\cdot)$ as the system state while $S^v(\cdot)$ as the system observation.

For fixed $x^v(\cdot)$, it is clear that (1.2) admits a unique solution. Set $Y^v(\cdot) = \log S^v(\cdot)$ and apply Itô's formula,

$$\begin{cases} dY^v(t) = \left(cx^v(t) - \frac{1}{2}\gamma^2(t) + a\right) dt + \gamma(t) d\tilde{W}(t), \\ Y^v(0) = \ln s_0. \end{cases} \tag{1.3}$$

We emphasize that the observation $Y^v(\cdot)$ depends on the premium policy $v(\cdot)$ implicitly through the cash-balance process $x^v(\cdot)$ while $v(\cdot)$ also depends on $Y^v(\cdot)$. This constitutes a "coupled circle" which makes the resulting optimal premium problem difficult to study. We will adopt a decoupling technique to overcome this difficulty. Note that the technique is inspired by Bensoussan [6]. Let us consider the state and observation equations

$$\begin{cases} dx_1(t) = (\delta(t)x_1(t) - b(t))dt + \sigma(t)dW(t), \\ dY_1(t) = \left(cx_1(t) + a - \frac{1}{2}\gamma^2(t)\right) dt + \gamma(t) d\tilde{W}(t), \\ x_1(0) = x_0, \quad Y_1(0) = \ln s_0 \end{cases} \tag{1.4}$$

and

$$\begin{cases} \dot{x}_2^v(t) = \delta(t)x_2^v(t) + v(t), \\ \dot{Y}_2^v(t) = cx_2^v(t), \\ x_2^v(0) = 0, \quad Y_2^v(0) = 0. \end{cases} \tag{1.5}$$

For any $v(\cdot) \in \mathscr{L}_{\mathbb{F}}^2(0,T;\mathbb{R})$, it is easy to check that $x_1(\cdot) + x_2^v(\cdot)$ and $Y_1(\cdot) + Y_2^v(\cdot)$ are the unique solutions of (1.1) and (1.3), i.e.,

$$x^v(\cdot) = x_1(\cdot) + x_2^v(\cdot), \qquad Y^v(\cdot) = Y_1(\cdot) + Y_2^v(\cdot).$$

Set $\mathscr{F}_t^{Y^v} = \{Y^v(s); 0 \le s \le t\}$ and $\mathscr{F}_t^{Y_1} = \sigma\{Y_1(s); 0 \le s \le t\}$. Since $x_1(\cdot)$ in (1.4) does not depend on $v(\cdot)$, neither does $\mathscr{F}_t^{Y_1}$. However, $\mathscr{F}_t^{Y^v}$ still depends on $v(\cdot)$ via $x^v(\cdot)$. To avoid the effect of $v(\cdot)$ on $\mathscr{F}_t^{Y^v}$, we introduce the following definition.

Definition 1.1. A premium policy $v : \Omega \times [0,T] \to \mathbb{R}$ is called admissible, if $v(t)$ is both $\mathscr{F}_t^{Y^v}$ and $\mathscr{F}_t^{Y_1}$-adapted with

$$\mathbb{E}\int_0^T v^4(t)dt < \infty$$

for each $0 \le t \le T$; for given $c_0 > 0$, (1.1) admits a unique solution $x^v(\cdot)$ satisfying

$$\mathbb{E}x^v(T) = c_0.$$

The set of all admissible policies is denoted by $\mathscr{U}_{ad}$.

From Definition 1.1, and the equations (1.4) and (1.5), it is easy to check that if $v(\cdot) \in \mathscr{U}_{ad}$, then $\mathscr{F}_t^{Y^v} \equiv \mathscr{F}_t^{Y_1}$. In fact, it is clear that $\mathscr{F}_t^{Y^v} \supseteq \mathscr{F}_t^{Y_1}$. On the other hand, if $v(\cdot) \in \mathscr{U}_{ad}$, from (1.5) we know that $x_2^v(t)$ is $\mathscr{F}_t^{Y_1}$-adapted, so is $Y_2^v(t)$. Then $Y^v(t) = Y_1(t) + Y_2^v(t)$ is $\mathscr{F}_t^{Y_1}$-adapted, i.e., $\mathscr{F}_t^{Y^v} \subseteq \mathscr{F}_t^{Y_1}$. This implies that $v(\cdot)$ depends on $\mathscr{F}_t^{Y^v}$ but $\mathscr{F}_t^{Y^v}$ is not affected by $v(\cdot)$. Thus the aforementioned "coupled circle" is uncoupled in the setup of Definition 1.1.

For any $v(\cdot) \in \mathscr{U}_{ad}$, we assume that the cost functional is

$$J(v(\cdot)) = \frac{1}{2}\mathbb{E}\left\{ \int_0^T e^{-\beta t}\left[L(t)(x^v(t) - A(t))^2 + N(t)v^2(t)\right] dt \right. \\ \left. + Me^{-\beta T}(x^v(T) - c_0)^2 \right\}, \tag{1.6}$$

where $L(\cdot) \geq 0$, $N(\cdot) > 0$, $N^{-1}(\cdot)$, $\delta(\cdot)$, $b(\cdot)$, $\sigma(\cdot)$, and $A(\cdot)$ are all deterministic and uniformly bounded on $[0, T]$; the terminal weight $M \geq 0$, and the discount factor $\beta > 0$; $A(\cdot)$ is some dynamic pre-set target; and $L(\cdot)$, $N(\cdot)$ and M are the weighting factors which make (1.6) more general and flexible to accommodate the preference of the policy-maker. Furthermore, we suppose that $A(\cdot)$ converges to c_0 as t goes to T, that is,

$$\lim_{t \to T} A(t) = c_0.$$

This assumption is reasonable and arises naturally by noting that $A(\cdot)$ actually represents the dynamic benchmark of $x(\cdot)$ evolving with time while c_0 specifies its terminal constraint when $t = T$. Then the optimal premium problem is

Problem (OP). To minimize (1.6) over $\mathscr{U}_{ad}$ subject to (1.1) and (1.3).

From (1.6), we can see that the insurance firm has three objectives: first, to minimize the "solvency risk," measured as the deviation between its cash-balance process, and some pre-set target $A(\cdot)$; second, to minimize the cost of premium policy over the whole time horizon; last, to minimize the terminal variance of cash-balance process satisfying a given constraint. Note that not one but all these three objectives must be achieved simultaneously. In other words, the concerns of the insurance business is to pay due benefits, but at the same time to prevent the cash balance from large deviations so as to stabilize the insurance schemes.

This is a partially observable LQ optimal control problem driven by SDE, which has been studied thoroughly by combining the backward separation approach with maximum principle and filtering for BSDE. We refer the reader to Chapter 5 for more details on the solution to this problem.

1.1.3 Risk Minimizing Problem

Consider an economic quantity which is governed by

$$\begin{cases} dx^v(t) = & (A(t)x^v(t) + B(t)v(t))\,dt + (C(t)v(t) + D(t))\,dW(t) \\ & + \left(\tilde{C}(t)v(t) + \tilde{D}(t)\right) d\tilde{W}^v(t), \\ x^v(0) = & x_0, \end{cases} \tag{1.7}$$

where $v(\cdot)$ is a control strategy, $A(\cdot)$, $B(\cdot)$, $C(\cdot)$, $D(\cdot)$, $\tilde{C}(\cdot)$, and $\tilde{D}(\cdot)$ are bounded and deterministic; $\tilde{W}^v(\cdot)$ is a stochastic process depending on $v(\cdot)$.

The economic quantity can be interpreted as the cash-balance, the wealth or the intrinsic value process in different fields of actuarial science, mathematical finance or mathematical economics, respectively. Once again, it is possible to observe partially $x^v(\cdot)$ via a factor model

$$\begin{cases} dY(t) = & \left(\frac{1}{\beta}\alpha(t, x^v(t)) - \frac{1}{2}\beta\right) dt + d\tilde{W}^v(t), \\ Y(0) = & 0, \end{cases} \tag{1.8}$$

where $\beta > 0$ is a constant, and α satisfies an assumption similar to h in (H1.6) below. $(W(\cdot), Y(\cdot))$ is a standard Brownian motion defined on $(\Omega, \mathscr{F}, (\mathscr{F}_t)_{0 \le t \le T}, \mathbb{P})$; and $\mathscr{F}_t$ is the natural filtration generated by $(W(t), Y(t))$.

A typical example of $Y(\cdot)$ in reality is the logarithm of the stock price $S(\cdot)$ related to $x(\cdot)$. In detail, set $S(\cdot) = s_0 e^{\beta Y(\cdot)}$ with a constant $s_0 > 0$. Obviously, the stock price $S(\cdot)$ is the information available to the policymaker. Moreover, it follows from Itô's formula that

$$\begin{cases} dS(t) = & S(t)\left(\alpha(t, x^v(t))dt + \beta d\tilde{W}^v(t)\right), \\ S(0) = & s_0, \end{cases} \qquad \text{(under } \mathbb{P}^v\text{)}$$

where $\mathbb{P}^v$ is a new probability measure under which $(W(\cdot), \tilde{W}^v(\cdot))$ is also a standard Brownian motion defined on $(\Omega, \mathscr{F}, (\mathscr{F}_t)_{0 \le t \le T}, \mathbb{P}^v)$. See, e.g., Section 1.2.2 below for more details.

Let $\rho(x^v(T))$ denote the risk of the economic quantity $x^v(T)$, where the risk measure $\rho(\cdot)$ is convex in the sense of Föllmer and Schied [23], and Frittelli and Rosazza–Gianin [24]. Define

$$\mathscr{F}_t^Y = \sigma\{Y(s); 0 \le s \le t\}.$$

We state the risk minimizing problem with partially observable information.

Problem (RM). Find an $\mathscr{F}_t^Y$-adapted $u(\cdot)$ such that

$$J(v(\cdot)) = \rho(x^v(T)) + \frac{1}{2}\mathbb{E}^v \int_0^T (v(t) - M(t))^2\, dt \tag{1.9}$$

is minimized subject to (1.7) and (1.8), where $M(\cdot)$ is a deterministic and bounded function taking values in $\mathbb{R}$, and is referred to as a dynamic benchmark. (1.9) implies that the policymaker wants not only to prevent the control strategy from large deviation, but also to minimize the risk of the economic quantity.

Recently, Rosazza–Gianin [71] established the relationship between the risk measure $\rho(\cdot)$ and the g-expectation $\mathscr{E}_g^v$:

$$\rho(x^v(T)) = \mathscr{E}_g^v[-x^v(T)] \equiv y^v(0),$$

where $(y^v(\cdot), z^v(\cdot), \tilde{z}^v(\cdot))$ is the unique solution to the following BSDE:

$$\begin{cases} -dy^v(t) = g(t, y^v(t), z^v(t), \tilde{z}^v(t))dt - z^v(t)dW(t) - \tilde{z}^v(t)d\tilde{W}^v(t), \\ \quad y^v(T) = -x^v(T), \end{cases} \tag{1.10}$$

with generator $g : [0,T] \times \mathbb{R}^3 \to \mathbb{R}$ satisfying $g(t,y,0,0) = 0$. Thus, in this situation Problem (RM) is equivalent to minimizing

$$J(v(\cdot)) = \mathbb{E}^v \left[y^v(0) + \frac{1}{2} \int_0^T (v(t) - M(t))^2 dt \right]$$

subject to (1.7), (1.8) and (1.10). This is a special optimal control problem driven by FBSDE with partially observable information.

1.2 Control Models

Motivated by the examples in the previous section, we introduce two optimal control models of FBSDE. If the available information at time t is only a sub-σ-field of the complete information $\mathscr{F}_t$, which does not depend on the control process, we call the control problem of FBSDE a partial information control model. If the available information is a noisy observation of the state up to time t, we call the control problem a partially observable control model. Comparing with the partial information model, the partially observable model is usually harder to study. An intrinsic difficulty comes from the circular dependence between the control and the observation. Namely, the control process must be adapted to the observation σ-field which depends on the control process itself. This results in the inapplicability of the variational method.

There are two ways to overcome this difficulty. One is the state decomposition technique for linear system, and the other is the measure transformation technique for nonlinear system. See, e.g., Bensoussan [6], Wang et al. [88, 89] for more details about this topic.

1.2.1 Partial Information Model

Consider a fully coupled FBSDE

$$\begin{cases} dx^v(t) = b(t,x^v(t),y^v(t),z^v(t),v(t))dt \\ \qquad\qquad +\sigma(t,x^v(t),y^v(t),z^v(t),v(t))dW(t), \\ -dy^v(t) = g(t,x^v(t),y^v(t),z^v(t),v(t))dt - z^v(t)dW(t), \\ x^v(0) = x_0, \quad y^v(T) = f(x^v(T)), \end{cases} \tag{1.11}$$

where $b, g : \Omega \times [0,T] \times \mathbb{R}^{n+n+n\times m+k} \to \mathbb{R}^n$, $\sigma : \Omega \times [0,T] \times \mathbb{R}^{n+n+n\times m+k} \to \mathbb{R}^{n\times m}$, $f : \Omega \times \mathbb{R}^n \to \mathbb{R}^n$ are continuous functions; $v(\cdot)$ is a control process; $W(\cdot)$ is an m-dimensional standard Brownian motion defined on $(\Omega, \mathscr{F}, (\mathscr{F}_t)_{0\le t\le T}, \mathbb{P})$; $\mathscr{F}_t$ is the natural filtration generated by $W(t)$; and $x_0 \in \mathbb{R}^n$.

Let U be a nonempty subset of $\mathbb{R}^k$, and let $\mathscr{G}_t$ be a sub-σ-field of $\mathscr{F}_t$, i.e.,

$$\mathscr{G}_t \subseteq \mathscr{F}_t, \quad t \in [0,T].$$

Here $\mathscr{G}_t$ could be the σ-field generated by some components of the underlying Brownian motion, or could be the time-delay information as follows:

$$\mathscr{G}_t = \mathscr{F}_{(t-\delta)^+}$$

with a given constant $\delta > 0$. The key difference between $\mathscr{G}_t$ here and $\mathscr{F}_t^Y$ in the partially observable model in the next subsection is that $\mathscr{G}_t$ does not depend on the control process while $\mathscr{F}_t^Y$ there does.

Define

$$\mathscr{U}_{ad} = \left\{ v(\cdot) : v(\cdot) \in \mathscr{L}^2_{\mathbb{G}}(0,T;\mathbb{R}^k), v(t) \in U, t \in [0,T] \right\}.$$

Every element of $\mathscr{U}_{ad}$ is called an admissible control.

Set

$$\lambda = \begin{pmatrix} x \\ y \\ z \end{pmatrix}, \bar{\lambda} = \begin{pmatrix} \bar{x} \\ \bar{y} \\ \bar{z} \end{pmatrix}, \text{ and } \Lambda(t,\tilde{\lambda}) = \begin{pmatrix} -g \\ b \\ \sigma \end{pmatrix} (t,\tilde{\lambda})$$

with $\tilde{\lambda} = \lambda, \bar{\lambda}$. We make the following assumptions.

(H1.1) *For any $(\lambda, v) \in \mathbb{R}^{n+n+n\times m} \times U$, $\Lambda(\cdot,\lambda)$ is an $\mathscr{F}_t$-adapted vector-valued process defined on $[0,T]$ with $\Lambda(\cdot,0) \in \mathscr{L}^2_{\mathbb{F}}(0,T;\mathbb{R}^{n+n+n\times m})$.*

(H1.2) *For any $t \in [0,T]$, the function χ ($\chi = b, \sigma, g, f$) is continuously differentiable in (x,y,z,v), and its partial derivatives χ_x, χ_y, χ_z, and χ_v are uniformly bounded.*

(H1.3) *There are nonnegative constants μ_1, μ_2, and μ_3 such that for any $t \in [0,T]$, $\lambda, \bar{\lambda} \in \mathbb{R}^{n+n+n\times m}$ and $v \in U$,*

$$\langle \Lambda(t,\lambda) - \Lambda(t,\bar{\lambda}), \lambda - \bar{\lambda} \rangle \le -\mu_1 |x-\bar{x}|^2 - \mu_2(|y-\bar{y}|^2 + |z-\bar{z}|^2),$$
$$\langle f(x) - f(\bar{x}), x - \bar{x} \rangle \ge \mu_3 |x-\bar{x}|^2,$$

with $\mu_1 + \mu_2 > 0$ and $\mu_2 + \mu_3 > 0$.

(H1.3)' *There are nonnegative constants μ_1, μ_2, and μ_3 such that for any $t \in [0,T]$, $\lambda, \bar{\lambda} \in \mathbb{R}^{n+n+n\times m}$ and $v \in U$,*

$$\langle A(t,\lambda) - A(t,\bar{\lambda}), \lambda - \bar{\lambda} \rangle \geq \mu_1 |x - \bar{x}|^2 + \mu_2 (|y - \bar{y}|^2 + |z - \bar{z}|^2),$$
$$\langle f(x) - f(\bar{x}), x - \bar{x} \rangle \leq -\mu_3 |x - \bar{x}|^2,$$

with $\mu_1 + \mu_2 > 0$ *and* $\mu_2 + \mu_3 > 0$.

Under (H1.1)–(H1.3) (or (H1.1), (H1.2), and (H1.3)'), for any $v(\cdot) \in \mathscr{U}_{ad}$, there is a unique solution $(x^v(\cdot), y^v(\cdot), z^v(\cdot)) \in \mathscr{L}^2_{\mathbb{F}}(0,T;\mathbb{R}^{n+n+n\times m})$ to (1.11). See, e.g., Theorem A.3 below or [28, 68] for more information.

Remark 1.1. The assumption (H1.3) or (H1.3)' is needed for the existence and uniqueness of the solution of the coupled FBSDE, and hence, not directly needed for the optimal control results. When the system is decoupled, or, when it reduced to a BSDE, the results we shall state in this book remain true without this monotonicity condition.

The cost functional is

$$J(v(\cdot)) = \mathbb{E}\left[\int_0^T l(t, x^v(t), y^v(t), z^v(t), v(t))dt + \psi(x^v(T)) + \phi(y^v(0))\right], \quad (1.12)$$

where $l : \Omega \times [0,T] \times \mathbb{R}^{n+n+n\times m+k} \to \mathbb{R}$, $\psi : \Omega \times \mathbb{R}^n \to \mathbb{R}$ and $\phi : \mathbb{R}^n \to \mathbb{R}$ satisfy

(H1.4) *for any* $t \in [0,T]$, l *is continuously differentiable in* (x,y,z), *and its partial derivatives* l_x, l_y, $l_z \in \mathscr{L}^2_{\mathbb{F}}(0,T)$. *Moreover, for any* $v(\cdot) \in \mathscr{U}_{ad}$, l *belongs to* $\mathscr{L}^1_{\mathbb{F}}(0,T)$;

(H1.5) ψ *and* ϕ *are continuously differentiable in* x *and* y, *respectively, and their derivatives* ψ_x *and* ϕ_y *grow linearly.*

We pose the optimal control problem of fully coupled FBSDE with partial information.

Problem A. Find an admissible control $u(\cdot) \in \mathscr{U}_{ad}$ satisfying

$$J(u(\cdot)) = \min_{v(\cdot)\in\mathscr{U}_{ad}} J(v(\cdot))$$

subject to (1.11). If such an admissible control exists, we call $u(\cdot)$ an optimal control and the corresponding state $(x(\cdot), y(\cdot), z(\cdot)) = (x^u(\cdot), y^u(\cdot), z^u(\cdot))$ an optimal state.

In [67], Peng formulated originally an optimal control problem of decoupled FBSDE with convex control domain. Since then, there has been an increasing research interest about Problem A and its various extensions. Let us list several papers which are most closely related to the current setup. Wu [99] generalized Peng [67] to the case that the state is governed by fully coupled FBSDE. Shi and Wu [75] and Wu [101] studied an optimal control problem of FBSDE with non-convex control domain. See also Huang and Shi [29], Ji and Wei [32], Li and Wei [46], Yong [107], Hu [26], and the references therein for other developments of this topic. Note that all of the works mentioned above are based on the assumption that the available information is complete, i.e.,

$$\mathscr{G}_t = \mathscr{F}_t, \quad t \in [0,T].$$

However, in reality it is possible that only partial information of $\mathscr{F}_t$ is obtainable. With the available information, Meng [54] obtained a maximum principle and a verification theorem for optimality of Problem A. Øksendal and Sulem [61] studied an optimal control problem of decoupled FBSDE with jump and diffusion. Wang and Xiao [90] extended [54, 61] slightly to the case of infinite time horizon. Arrow's verification theorem for optimality was derived. Several practical examples arising from finance and economics were solved explicitly by using the verification theorem and the filtering of FBSDE. Note that since neither [54] nor [61] contains infinite horizon and filtering, they are different from [90].

The following two important special cases of Problem A will be studied in more detail.

Case A1. Neither the state equation (1.11) nor the cost functional (1.12) contains $(y^v(\cdot), z^v(\cdot))$, i.e., for any $(\omega, t, x, y, z, v) \in \Omega \times [0, T] \times \mathbb{R}^{n+n+n\times m} \times U$,

$$\begin{gathered} b(t,x,y,z,v) \equiv b(t,x,v), \quad \sigma(t,x,y,z,v) \equiv \sigma(t,x,v), \\ l(t,x,y,z,v) \equiv l(t,x,v), \quad g(t,x,y,z,v) \equiv 0, \\ f(x) \equiv 0, \quad \phi(y) \equiv 0. \end{gathered}$$

In this case, Problem A reduces to minimize

$$J(v(\cdot)) = \mathbb{E}\left[\int_0^T l(t, x^v(t), v(t))dt + \psi(x^v(T))\right]$$

over $\mathscr{U}_{ad}$ subject to

$$\begin{cases} dx^v(t) = b(t, x^v(t), v(t))dt + \sigma(t, x^v(t), v(t))dW(t), \\ x^v(0) = x_0. \end{cases}$$

This case has been well studied under various settings. We refer the reader to Yong and Zhou [109] and the references therein for more details toward maximum principle, dynamic programming, LQ control, and their applications to mathematical finance. See also Hu and Øksendal [27], Wang and Wu [87], etc. for other developments about LQ control and mean-variance hedging with partial information.

Case A2. Both state equation (1.11) and cost functional (1.12) are independent of $x^v(\cdot)$, i.e., for any $(\omega, t, x, y, z, v) \in \Omega \times [0, T] \times \mathbb{R}^{n+n+n\times m} \times U$,

$$\begin{gathered} b(t,x,y,z,v) \equiv 0, \quad \sigma(t,x,y,z,v) \equiv 0, \quad \psi(x) \equiv 0, \quad f(x) \equiv \xi, \\ l(t,x,y,z,v) \equiv l(t,y,z,v), \quad g(t,x,y,z,v) \equiv g(t,y,z,v). \end{gathered}$$

With these restrictions, we get an important class of optimal control problem with partial information. To find an admissible control $v(\cdot) \in \mathscr{U}_{ad}$ such that

$$J(v(\cdot)) = \mathbb{E}\left[\int_0^T l(t, y^v(t), z^v(t), v(t))dt + \phi(y^v(0))\right]$$

is minimized, subject to

$$\begin{cases} -dy^v(t) = g(t,y^v(t),z^v(t),v(t))dt - z^v(t)dW(t), \\ y^v(T) = \xi. \end{cases}$$

Note that since the state is governed by a BSDE rather than an SDE, this problem is essentially different from that of Case 1. Lim and Zhou [48] solved an LQ version of Case 2 with complete information. Huang et al. [31] studied the optimal control problem of Case 2. Via the convex variation, a maximum principle for optimality was derived and was used to study an LQ example with partial information. Note that the traditional separation principle is not applicable for the LQ example. Instead, a new backward separation approach was introduced. Combining it with the filtering of FBSDE, an optimal feedback control for the LQ example was obtained explicitly. We emphasize that the filtering of FBSDE is new and arises naturally from the deductions of the optimal control, and thus, it should not be viewed as an artificial extension of the classical filtering. Along this line, Wang and Yu [94] investigated a nonzero sum differential game of BSDE. Maximum principle and verification theorem for equilibrium point were derived. An approach to prove the existence and uniqueness of equilibrium point was found. These theoretical results obtained in [94] were then used to solve an LQ game and a stock differential game both in the incomplete information case. In a recent study of Wang et al. [92], the game in [94] was extended to the case of asymmetric information.

1.2.2 Partially Observable Model

Consider a controlled FBSDE

$$\begin{cases} dx^v(t) = b(t,x^v(t),v(t))dt + \sigma(t,x^v(t),v(t))dW(t) \\ \qquad\qquad + \tilde{\sigma}(t,x^v(t),v(t))d\tilde{W}^v(t), \\ -dy^v(t) = g(t,x^v(t),y^v(t),z^v(t),\tilde{z}^v(t),v(t))dt \\ \qquad\qquad - z^v(t)dW(t) - \tilde{z}^v(t)dY(t), \\ x^v(0) = x_0, \quad y^v(T) = f(x^v(T)). \end{cases} \tag{1.13}$$

Here $v(\cdot)$ is a control process taking values in a nonempty subset U of $\mathbb{R}^r$; $b : [0,T] \times \mathbb{R}^n \times U \to \mathbb{R}^n$, $\sigma : [0,T] \times \mathbb{R}^n \times U \to \mathbb{R}^{n\times k}$, $\tilde{\sigma} : [0,T] \times \mathbb{R}^n \times U \to \mathbb{R}^{n\times \tilde{k}}$, $g : [0,T] \times \mathbb{R}^{n+m+m\times k+m\times \tilde{k}} \times U \to \mathbb{R}^m$, and $f : \mathbb{R}^n \to \mathbb{R}^m$ are given continuous mappings; $x_0 \in \mathbb{R}^n$ is the initial state; $(W(\cdot),Y(\cdot))$ is a $(k+\tilde{k})$-dimensional standard Brownian motion defined on $(\Omega, \mathscr{F}, (\mathscr{F}_t)_{0\le t\le T}, \mathbb{P})$; and $\tilde{W}^v(\cdot)$ is a $\tilde{k}$-dimensional stochastic process depending on $v(\cdot)$. It should be noted that in this case the control $v(\cdot)$ enters into the diffusion coefficients σ and $\tilde{\sigma}$.

Suppose that $(x^v(\cdot),y^v(\cdot),z^v(\cdot),\tilde{z}^v(\cdot))$ can only be observed partially via $Y(\cdot)$, which is governed by the SDE

$$\begin{cases} dY(t) = h(t,x^v(t))dt + d\tilde{W}^v(t), \\ Y(0) = 0, \end{cases} \tag{1.14}$$

where $h : [0,T] \times \mathbb{R}^n \to \mathbb{R}^{\tilde{k}}$ is a given continuous mapping. Let

$$\mathscr{F}_t^Y = \sigma\{Y(s); 0 \le s \le t\}.$$

Definition 1.2. A control process $v : \Omega \times [0,T] \to U$ is called admissible, if $v(t)$ is $\mathscr{F}_t^Y$-adapted such that

$$\sup_{0 \le t \le T} \mathbb{E}|v(t)|^8 < \infty.$$

The set of all admissible controls is denoted by $\mathscr{U}_{ad}$.

(H1.6) *The functions b, σ, $\tilde{\sigma}$, f, g and h are continuously differentiable with respect to $x, y, z, \tilde{z}, v$, respectively, and their partial derivatives b_x, b_v, σ_x, σ_v, $\tilde{\sigma}_x$, $\tilde{\sigma}_v$, h_x, g_x, g_y, g_z, $g_{\tilde{z}}$, g_v, and f_x are uniformly bounded. Moreover, for any $(t,x,v) \in [0,T] \times \mathbb{R}^n \times U$, there is a constant C such that*

$$|\tilde{\sigma}(t,x,v)| + |h(t,x)| \le C.$$

Note that the boundedness of h can be relaxed to linear growth with respect to x under some restrictive assumptions about control processes. See, e.g., Wang et al. [88] for more details about this generalization.

Inserting (1.14) into (1.13), we have

$$\begin{cases} dx^v(t) = & [b(t,x^v(t),v(t)) - \tilde{\sigma}(t,x^v(t),v(t))h(t,x^v(t))]\,dt \\ & +\sigma(t,x^v(t),v(t))dW(t) + \tilde{\sigma}(t,x^v(t),v(t))dY(t), \\ -dy^v(t) = & g(t,x^v(t),y^v(t),z^v(t),\tilde{z}^v(t),v(t))dt \\ & -z^v(t)dW(t) - \tilde{z}^v(t)dY(t), \\ x^v(0) = x_0, & y^v(T) = f(x^v(T)). \end{cases} \tag{1.15}$$

Since FBSDE (1.15) is decoupled, for any $v(\cdot) \in \mathscr{U}_{ad}$, it admits a unique solution under (H1.6), which is denoted by

$$(x^v(\cdot), y^v(\cdot), z^v(\cdot), \tilde{z}^v(\cdot)) \in \mathscr{L}_{\mathbb{F}}^2(0,T;\mathbb{R}^{n+m+m\times k+m\times \tilde{k}}).$$

Let us introduce a process

$$Z^v(t) = \exp\left\{\int_0^t h^\top(s,x^v(s))dY(s) - \frac{1}{2}\int_0^t |h(s,x^v(s))|^2 ds\right\},$$

which is the solution to the SDE

$$\begin{cases} dZ^v(t) = & Z^v(t)h^\top(t,x^v(t))dY(t), \\ Z^v(0) = & 1. \end{cases} \tag{1.16}$$

Under (H1.6), $Z^v(\cdot)$ is an $((\mathscr{F}_t)_{0 \le t \le T}, \mathbb{P})$-martingale. We can then define a new probability measure $\mathbb{P}^v$ such that for any t,

$$d\mathbb{P}^v = Z^v(t)d\mathbb{P} \quad \text{on } \mathscr{F}_t. \tag{1.17}$$

According to Girsanov's theorem and (1.14), $(W(\cdot),\tilde{W}^v(\cdot))$ is a $(k+\tilde{k})$-dimensional standard Brownian motion defined on $(\Omega,\mathscr{F},(\mathscr{F}_t)_{0\le t\le T},\mathbb{P}^v)$.

The associated cost functional is given by

$$J(v(\cdot))=\mathbb{E}^v\left[\int_0^T l(t,x^v(t),y^v(t),z^v(t),\tilde{z}^v(t),v(t))dt+\phi(x^v(T))+\gamma(y^v(0))\right], \tag{1.18}$$

where $\mathbb{E}^v$ stands for the mathematical expectation on $(\Omega,\mathscr{F},(\mathscr{F}_t)_{0\le t\le T},\mathbb{P}^v)$, and the following assumption on l, ϕ, and γ will be needed.

(H1.7) *The mappings* $l:[0,T]\times\mathbb{R}^{n+m+m\times k+m\times\tilde{k}}\times U\to\mathbb{R}$, $\phi:\mathbb{R}^n\to\mathbb{R}$ *and* $\gamma:\mathbb{R}^m\to\mathbb{R}$ *are continuously differentiable with respect to* $x,y,z,\tilde{z},v$, *respectively, and satisfy*

$$\mathbb{E}^v\left[\int_0^T |l(t,x^v(t),y^v(t),z^v(t),\tilde{z}^v(t),v(t))|dt+|\phi(x^v(T))|+|\gamma(y^v(0))|\right]<\infty, \tag{1.19}$$

i.e., $l(\cdot,x^v(\cdot),y^v(\cdot),z^v(\cdot),\tilde{z}^v(\cdot),v(\cdot))\in\mathscr{L}^1_{\mathscr{F}}(0,T;\mathbb{R})$ *and* $\phi\in\mathscr{L}^1_{\mathscr{F}}(\Omega;\mathbb{R})$.

Now we pose an optimal control problem of FBSDE with partially observable information.

Problem B. Seek an admissible control $u(\cdot)$ such that

$$J(u(\cdot))=\min_{v(\cdot)\in\mathscr{U}_{ad}} J(v(\cdot)) \tag{1.20}$$

subject to (1.13) and (1.14). If such an admissible control exists, we call $u(\cdot)$ an optimal control and the corresponding state

$$(x(\cdot),y(\cdot),z(\cdot),\tilde{z}(\cdot))=(x^u(\cdot),y^u(\cdot),z^u(\cdot),\tilde{z}^u(\cdot))$$

an optimal state.

Obviously, Problem B covers the foregoing Problem (RM) as a special case. Although there exist several papers related to Problem B, generally speaking, it is a new and unexplored topic. The early study about this topic can be traced back to Wu [100], where the drift coefficient of the observation equation is bounded, and the state noise is independent of the observation noise. Using the backward separation approach and the traditional techniques in the complete information case, a maximum principle for optimality was expressed by the conditional expectation of a function with respect to an observable filtration. Furthermore, the filtering of the adjoint processes was obtained, and was used to describe an optimal control. Along this line, there are a few works including Wang and Wu [85] and Xiao [103]. We emphasize that the drift coefficients of the observation equations in [100, 85, 103] are uniformly bounded with respect to (t,x,v). The assumption simplifies the computations of this topic. However, it excludes some important applications in reality. Very recently, Wang et al. [88] improved [100]. They assumed that h grows linearly with respect to x, but the diffusion coefficient σ is uniformly bounded in (t,x,v); moreover, the collection of admissible controls is a subset of $\mathscr{U}_{ad}$. Combining high-order moment estimates of the adjoint processes of $(x(\cdot),y(\cdot),z(\cdot),Z(\cdot))$ with an approximation method by bounded and smooth functions, a maximum principle for optimality was established.

Note that the approximation method introduced in [88] is invalid to address Problem B. A main reason is as follows. With the assumption of h being linear with respect to x, the adjoint process related to $(x(\cdot),y(\cdot),z(\cdot),Z(\cdot))$ satisfies an FBSDE with square-integrable stochastic coefficients. In general, it is difficult to study the solvability and to obtain the high-order moment estimate of the FBSDE. Then the relevant variational inequality for Problem B cannot be derived under the new probability $\mathbb{P}^v$. The most recent attempt to address Problem B was done by Wang et al. [89], where a decomposition approach was introduced to study an LQ case of Problem B. Combining the decomposition approach with the backward separation approach and filtering, optimality conditions and a feedback representation of the optimal control were derived. As an application of the optimality conditions, a generalized recursive utility problem from financial market was solved explicitly. Other cases of Problem B are worthy of further study in the future.

The following important special case of Problem B is also very interesting.

Case B1. To minimize

$$J(v(\cdot)) = \mathbb{E}^v\left[\int_0^T l(t,x^v(t),v(t))dt + \phi(x^v(T))\right]$$

over $v(\cdot) \in \mathscr{U}_{ad}$, subject to

$$\begin{cases} dx^v(t) = b(t,x^v(t),v(t))dt + \sigma(t,x^v(t),v(t))dW(t) + \tilde{\sigma}(t,x^v(t),v(t))d\tilde{W}^v(t), \\ x^v(0) = x_0 \end{cases}$$

and (1.14).

This problem has been well understood so far in the case that the diffusion coefficient is independent of the control. We refer to Bensoussan [6] and the references therein for early studies on maximum principle and LQ control. After that, many versions of the maximum principles were derived under various settings, say, the control domain is a non-convex set, and/or the state noise is correlated to the observation noise, and/or the cost functional is a risk sensitive one. See, e.g., Zhang [114], Li and Tang [45], Tang [78], and Wang and Wu [86] for more details.

1.3 An Overview

In this section, we provide an outline of the results which will be studied in this book. In Chapter 2, we present some filtering equations for BSDEs, which are different from the traditional ones. Chapters 3 and 4 are devoted to addressing Problem A and Problem B, respectively. In Chapter 5, several LQ examples with incomplete information are solved explicitly. Finally, we will provide an appendix to introduce the basic material of BSDE and FBSDE which will be used in the rest of this book.

We refer the reader to Ma and Yong [52], Yong and Zhou [109], and Ma et al. [51] for detailed treatments of this topic.

We now sketch the results of Chapters 2–5. In Chapter 2, we will introduce some filtering results for BSDEs. For simplicity of notation, we confine ourselves to the 1-dimensional case, i.e., $m = k = \tilde{k} = 1$. Also, we adopt the notations

$$\hat{y}(t) = \mathbb{E}[y(t)|\mathscr{F}_t^Y], \quad \hat{z}(t) = \mathbb{E}[z(t)|\mathscr{F}_t^Y], \quad \hat{h}(t) = \mathbb{E}[h(\omega,t)|\mathscr{F}_t^Y],$$

$$\hat{g}(t) = \mathbb{E}[g(\omega,t,y(t),z_1(t),z_2(t))|\mathscr{F}_t^Y], \quad \widehat{(yh)}(t) = \mathbb{E}[y(t)h(\omega,t)|\mathscr{F}_t^Y].$$

Suppose that the state process $(y(\cdot),z_1(\cdot),z_2(\cdot))$ is governed by

$$y(t) = \xi + \int_t^T g(\omega,s,y(s),z_1(s),z_2(s))ds - \int_t^T z_1(s)dW_1(s) - \int_t^T z_2(s)d\tilde{W}_2(s),$$

whose noisy observation $Y(\cdot)$ satisfies

$$Y(t) = \int_0^t h(\omega,s)ds + \int_0^t f(s)dW(s).$$

Here $g : \Omega \times [0,T] \times \mathbb{R}^3 \to \mathbb{R}$, $h : \Omega \times [0,T] \to \mathbb{R}$ and $f : [0,T] \to \mathbb{R}$ are given mappings; ξ is a square-integrable random variable; $(W_1(\cdot),W_2(\cdot))$ is a 2-dimensional standard Brownian motion defined on $(\Omega,\mathscr{F},(\mathscr{F}_t)_{0\le t\le T},\mathbb{P})$; and $\mathscr{F}_t$ is the natural filtration generated by $(W_1(\cdot),W_2(\cdot))$. According to Malliavin calculus, $z_1(\cdot)$ and $z_2(\cdot)$ are equal to the Malliavin derivatives of $y(\cdot)$ with respect to $W_1(\cdot)$ and $W_2(\cdot)$, respectively. We are interested in computing the filtering $\hat{y}(t)$ of $y(\cdot)$ based on the observable information

$$\mathscr{F}_t^Y = \sigma\{Y(s); 0 \le s \le t\}.$$

Theorem 1.1. *Under suitable conditions about g, h, and f, the filtering $\hat{y}(\cdot)$ satisfies a BSDE*

$$\hat{y}(t) = \mathbb{E}\left[\xi|\mathscr{F}_T^Y\right] + \int_t^T \hat{g}(s)ds - \int_t^T \left\{\hat{z}_1(s) + \frac{1}{f(s)}\left[\widehat{(yh)}(s) - \hat{y}(s)\hat{h}(s)\right]\right\}d\hat{W}(s), \tag{1.21}$$

where

$$\hat{W}(t) = \int_0^t \frac{1}{f(s)}\left(dY(s) - \hat{h}(s)ds\right) \tag{1.22}$$

called the innovation process, is a 1-dimensional standard Brownian motion defined on $(\Omega,\mathscr{F}^Y,(\mathscr{F}_t^Y)_{0\le t\le T},\mathbb{P})$.

With some detailed assumptions, equation (1.21) is reduced to a BSDE rather than an SDE, which shows the difference from the traditional filtering theory. Here is an interesting example. Assume that g and h satisfy

$$\hat{g}(t) = g(t,\hat{y}(t),\hat{z}_1(t)) \text{ and } \widehat{(yh)}(t) = \hat{y}(t)\hat{h}(t),$$

respectively. Then the filtering $(\hat{y}(t), \hat{z}_1(t))$ of $(y(t), z_1(t))$ with respect to $\mathscr{F}_t^Y$ is the solution to the standard BSDE

$$\hat{y}(t) = \mathbb{E}\left[\xi|\mathscr{F}_T^Y\right] + \int_t^T g(s, \hat{y}(s), \hat{z}_1(s))ds - \int_t^T \hat{z}_1(s)d\hat{W}(s),$$

where $\hat{W}(\cdot)$ is defined by (1.22). One more example can be found in Chapter 2, where the Kalman-Bucy filtering for a class of FBSDEs was obtained.

In Chapter 3, we aim at deriving a stochastic maximum principle and a verification theorem for optimality of Problem A. To obtain these results, we first introduce the following Hamiltonian function

$$\begin{aligned} H(t,x,y,z,v;p,q,k) = &\langle q, b(t,x,y,z,v)\rangle + \langle k, \sigma(t,x,y,z,v)\rangle \\ &- \langle p, g(t,x,y,z,v)\rangle + l(t,x,y,z,v). \end{aligned}$$

For simplicity of notation, we denote

$$H^u(t) = H(t, x(t), y(t), z(t), u(t); p(t), q(t), k(t)),$$

for $u \in \mathscr{U}_{ad}$. Then, we define the adjoint equation

$$\begin{cases} dp(t) = -H_y^u(t)dt - H_z^u(t)dW(t), \\ -dq(t) = H_x^u(t)dt - k(t)dW(t), \\ p(0) = -\phi_y(y(0)), \quad q(T) = \psi_x(x(T)) - f_x^\top(x(T))p(T). \end{cases} \tag{1.23}$$

In this situation, the resulting deduction is similar to the case of complete information. Using the convex variation and the dual technique, we establish the maximum principle for optimality of Problem A.

Theorem 1.2. *Under assumptions (H1.1)–(H1.5), if $u(\cdot)$ is an optimal control, then (1.23) admits a unique solution $(p(\cdot), q(\cdot), k(\cdot)) \in \mathscr{L}_{\mathbb{F}}^2(0,T;\mathbb{R}^{n+m+m\times k})$ such that for any $v \in U$ we have*

$$\langle \mathbb{E}\left[H_v(t)|\mathscr{G}_t\right], v - u(t)\rangle \geq 0. \tag{1.24}$$

With additional assumptions, the minimum condition given below is sufficient for the optimality.

Theorem 1.3. *Let (H1.1)–(H1.5) hold. Assume that*

- *(Terminal condition) for any $v(\cdot) \in \mathscr{U}_{ad}$, $y^v(T) = Ax^v(T)$, $A \in \mathbb{R}^{n\times n}$;*
- *(Minimum condition) for any $t \in [0,T]$,*

$$\mathbb{E}[H(t)|\mathscr{G}_t] = \min_{v\in U} \mathbb{E}[H(t, x(t), y(t), z(t), v; p(t), q(t), k(t))|\mathscr{G}_t].$$

- *(Convexity) for any $t \in [0,T]$, the Hamiltonian function*

$$H(t,x,y,z,v;p(t),q(t),k(t))$$

is convex in (x,y,z,v), and ψ and ϕ are convex in x and y.

Then $u(\cdot)$ is an optimal control of Problem A.

Remark 1.2. (1) The minimum condition (1.24) in Theorem 1.2 shows that we need to compute the conditional expectation of the solution to FBSDE in order to obtain $u(\cdot)$. This is one of the motivations that we study the filtering theory of BSDE in Chapter 2.

(2) If $\mathscr{G}_t = \mathscr{F}_t, t \in [0,T]$, then Theorem 1.3 is reduced to a complete information version. Moreover, the convexity condition of

$$H(t,x,y,z,v;p(t),q(t),k(t))$$

with respect to (x,y,z,v) can be weakened to the assumption that

$$\tilde{H}(t,x,y,z) = \min_{v\in U} H(t,x,y,z,v;p(t),q(t),k(t))$$

exists and is convex in (x,y,z). With this assumption, Arrow's sufficient condition for optimality can be derived, and can be used to solve an LQ example. We refer the reader to Chapter 3 below for more details.

In Chapter 4, we will derive two versions of the stochastic maximum principle for Problem B. For simplicity of notation, we only consider the 1-dimensional case. We introduce the adjoint equations

$$\begin{cases} -dP(t) = & l(t,x(t),y(t),z(t),\tilde{z}(t),u(t))dt - Q(t)dW(t) - \tilde{Q}(t)d\tilde{W}^u(t), \\ P(T) = & \phi(x(T)), \end{cases} \tag{1.25}$$

and

$$\begin{cases} dp(t) = & [g_y(t,x(t),y(t),z(t),\tilde{z}(t),u(t))p(t) \\ & \quad -l_y(t,x(t),y(t),z(t),\tilde{z}(t),u(t))]dt \\ & +[g_z(t,x(t),y(t),z(t),\tilde{z}(t),u(t))p(t) \\ & \quad -l_z(t,x(t),y(t),z(t),\tilde{z}(t),u(t))]dW(t) \\ & +[(g_{\tilde{z}}(t,x(t),y(t),z(t),\tilde{z}(t),u(t)) - h(t,x(t)))\,p(t) \\ & \quad -l_{\tilde{z}}(t,x(t),y(t),z(t),\tilde{z}(t),u(t))]\,d\tilde{W}^u(t), \\ -dq(t) = & \{[b_x(t,x(t),u(t)) - \tilde{\sigma}(t,x(t),u(t))h_x(t,x(t))]\,q(t) \\ & \quad +\sigma_x(t,x(t),u(t))k(t) + \tilde{\sigma}_x(t,x(t),u(t))\tilde{k}(t) \\ & \quad +h_x(t,x(t))\tilde{Q}(t) - g_x(t,x(t),y(t),z(t),\tilde{z}(t),u(t))p(t) \\ & \quad +l_x(t,x(t),y(t),z(t),\tilde{z}(t),u(t))\}dt \\ & -k(t)dW(t) - \tilde{k}(t)d\tilde{W}^u(t), \\ p(0) = & -\gamma_y(y(0)), \quad q(T) = -f_x(x(T))p(T) + \phi_x(x(T)). \end{cases} \tag{1.26}$$

Note that because of the dependency of (1.13) on $\tilde{W}^v(\cdot)$, (1.25) and (1.26) are dramatically different from the classical ones. Moreover, (1.25) is used to treat the terms induced by partially observable information, which is unnecessary in the cases of Peng [66], Øksendal and Sulem [61], Wu [100], and Yong [107].

Combining the Girsanov transformation with the convex variation and Taylor's expansion, we obtain the first version of the stochastic maximum principle for Problem B.

Theorem 1.4. *Let Assumptions (H1.6) and (H1.7) hold. Suppose that for any* $v(\cdot) \in \mathscr{U}_{ad}$*, the functions* ϕ*,* $\phi_x \in \mathscr{L}^2_{\mathscr{F}_T}(\Omega;\mathbb{R})$*,* l*,* l_x*,* l_y*,* l_z*,* $l_{\tilde{z}}$*,* $l_v \in \mathscr{L}^2_{\mathbb{F}}(0,T;\mathbb{R})$*. Furthermore, suppose that (1.25) and (1.26) admit unique solutions* $(P(\cdot),Q(\cdot),\tilde{Q}(\cdot)) \in \mathscr{L}^2_{\mathbb{F}}(0,T;\mathbb{R}^3)$ *and* $(p(\cdot),q(\cdot),k(\cdot),\tilde{k}(\cdot)) \in \mathscr{L}^2_{\mathbb{F}}(0,T;\mathbb{R}^4)$*, respectively. If* $u(\cdot)$ *is an optimal control of Problem B, then for any* $v \in U$ *we have*

$$\mathbb{E}^u\left[H_v(t,x(t),y(t),z(t),\tilde{z}(t),u(t);p(t),q(t),k(t),\tilde{k}(t),\tilde{Q}(t))(v-u(t))|\mathscr{F}^Y_t\right] \geq 0,$$

where the Hamiltonian function H is defined by

$$\begin{aligned} & H(t,x,y,z,\tilde{z},v;p,q,k,\tilde{k},\tilde{Q}) \\ = \; & b(t,x,v)q+\sigma(t,x,v)k+\tilde{\sigma}(t,x,v)\tilde{k}+h(t,x)\tilde{Q} \\ & -(g(t,x,y,z,\tilde{z},v)-h(t,x)\tilde{z})\,p+l(t,x,y,z,\tilde{z},v). \end{aligned}$$

The maximum principle depends explicitly on (1.25) and (1.26), which are usually very hard to solve. For this reason, we will introduce a method based on Malliavin calculus. The main feature of this method is that the *backward* adjoint process $(q(\cdot),k(\cdot),\tilde{k}(\cdot))$ in Theorem 1.4 is described almost directly via (forward) SDEs, while $(P(\cdot),Q(\cdot),\tilde{Q}(\cdot))$ is unnecessary.

We now define the new adjoint processes $\bar{q}(\cdot)$, $\bar{k}(\cdot)$, and $\bar{\tilde{k}}(\cdot)$ as follows:

$$\bar{q}(t) = \Sigma(t) + \int_t^T \Psi(t,s)ds, \quad \bar{k}(t) = D_t^{(W)}\bar{q}(t), \quad \bar{\tilde{k}}(t) = D_t^{(\tilde{W})}\bar{q}(t)$$

with

$$\Psi(t,s) = H_x(s)\Phi(t,s)$$

and

$$\Sigma(t) = \phi_x(x(T)) - f_x(x(T))\bar{p}(T) + \int_t^T l_x(s,x,y,z,\tilde{z},u)ds,$$

where

$$\begin{aligned} \Phi(t,s) = \; & \exp\left\{\textstyle\int_t^s \left[b_x(r,x,u)-\tilde{\sigma}(r,x,u)h_x(r,x)-\frac{1}{2}\left(\sigma_x^2(r,x,u)+\tilde{\sigma}_x^2(r,x,u)\right)\right]dr \right. \\ & \left. +\textstyle\int_t^s \sigma_x(r,x,u)dW(r)+\int_t^s \tilde{\sigma}_x(r,x,u)d\tilde{W}(r)\right\}, \end{aligned}$$

$$\begin{aligned} H_x(t) = \; & \Sigma(t)\left(b_x(t,x,u)-\tilde{\sigma}(t,x,u)h_x(t,x)\right)+\sigma_x(t,x,u)D_t^{(W)}\Sigma(t) \\ & +\tilde{\sigma}_x(t,x,u)D_t^{(\tilde{W})}\Sigma(t)+h_x(t,x)D_t^{(\tilde{W})}\Pi(t)-g_x(t,x,y,z,\tilde{z},u)\bar{p}(t), \end{aligned}$$

$$\Pi(t) = \phi(x(T)) + \int_t^T l(s,x,y,z,\tilde{z},u)ds,$$

and the adjoint process $\bar{p}(\cdot)$ satisfies

$$\begin{cases} d\bar{p}(t) = & \Big[g_y(t,x(t),y(t),z(t),\tilde{z}(t),u(t))\bar{p}(t) \\ & \quad -l_y(t,x(t),y(t),z(t),\tilde{z}(t),u(t))\Big]dt \\ & +\Big[g_z(t,x(t),y(t),z(t),\tilde{z}(t),u(t))\bar{p}(t) \\ & \quad -l_z(t,x(t),y(t),z(t),\tilde{z}(t),u(t))\Big]dW(t) \\ & +\left[(g_{\tilde{z}}(t,x(t),y(t),z(t),\tilde{z}(t),u(t))-h(t,x(t)))\,\bar{p}(t)\right. \\ & \quad \left.-l_{\tilde{z}}(t,x(t),y(t),z(t),\tilde{z}(t),u(t))\right]d\tilde{W}^u(t), \\ \bar{p}(0) = & -\gamma_y(y(0)). \end{cases} \tag{1.27}$$

Note that (1.27) is the same as the forward equation for $p(\cdot)$ in (1.26). With the adjoint processes and the convex variation, we obtain the second version of the stochastic maximum principle for the optimality of Problem B.

Theorem 1.5. *Let (H1.6) and (H1.7) hold. Suppose that (1.27) admits a unique solution $\bar{p}(\cdot) \in \mathscr{L}^2_{\mathbb{F}}(0,T;\mathbb{D}_{1,2})$. Assume that ϕ, $\phi_x \in \mathbb{D}_{1,2}$, l, l_x, and $\Psi(t,s)$ are in $\mathbb{L}_{1,2}(\mathbb{R})$ for all $0 \le t \le s \le T$. If $u(\cdot)$ is an optimal control, then for any $v \in U$ we have*

$$\mathbb{E}^u\left[\bar{H}_v(t,x(t),y(t),z(t),\tilde{z}(t),u(t);\bar{p}(t),\bar{q}(t),\bar{k}(t),\bar{\bar{k}}(t))(v-u(t))\Big|\mathscr{F}_t^Y\right] \ge 0,$$

where $\bar{H}_v$ is defined by

$$\begin{aligned} \bar{H}_v(t,x,y,z,\tilde{z},v;\bar{p},\bar{q},\bar{k},\bar{\bar{k}}) = \; & b_v(t,x,v)\bar{q}+\sigma_v(t,x,v)\bar{k}+\tilde{\sigma}_v(t,x,v)\bar{\bar{k}} \\ & -g_v(t,x,y,z,\tilde{z},v)\bar{p}+l_v(t,x,y,z,\tilde{z},v). \end{aligned}$$

Combining Theorem 1.1 with Theorem 1.4 or Theorem 1.5, we will solve several financial problems explicitly.

Finally, in Chapter 6, we will present two typical models with incomplete information. The first example is related to an LQ model of FBSDE with partially observable information. This model generalizes Problem B in the sense that the drift coefficient in the observation equation is linear with respect to the state, and thus, this model covers Problem (OP) in Section 1.1 as a special case, although a state constraint is included there.

Define the processes $\left(x^0(\cdot),y^0(\cdot),z^0(\cdot),\tilde{z}^0(\cdot)\right)$, and $Y^0(\cdot)$ by

$$\begin{cases} dx^0(t) = & a(t)x^0(t)dt+c(t)dW(t)+\tilde{c}(t)d\tilde{W}(t), \\ -dy^0(t) = & \left(A(t)x^0(t)+B(t)y^0(t)+C(t)z^0(t)+\tilde{C}(t)\tilde{z}^0(t)\right)dt \\ & -z^0(t)dW(t)-\tilde{z}^0(t)d\tilde{W}(t), \\ x^0(0) = & x_0, \quad y^0(T)=Fx^0(T), \end{cases} \tag{1.28}$$

and

$$\begin{cases} dY^0(t) = & f(t)x^0(t)dt+h(t)dw(t), \\ Y^0(0) = & 0. \end{cases} \tag{1.29}$$

Let $v(\cdot) \in \mathscr{L}^2_{\mathbb{F}}(0,T;\mathbb{R})$ be a control process. Define $\big(x^1(\cdot), y^1(\cdot), z^1(\cdot), \tilde{z}^1(\cdot)\big)$, and $Y^1(\cdot)$ by

$$\begin{cases} \dot{x}^1(t) = a(t)x^1(t) + b(t)v(t) + \tilde{b}(t), \\ -dy^1(t) = \big(A(t)x^1(t) + B(t)y^1(t) + C(t)z^1(t) + \tilde{C}(t)\tilde{z}^1(t) \\ \qquad\qquad + D(t)v(t) + \tilde{D}(t)\big)\, dt - z^1(t)dW(t) - \tilde{z}^1(t)d\tilde{W}(t), \\ x^1(0) = 0, \quad y^1(T) = Fx^1(T) + G \end{cases} \tag{1.30}$$

and

$$\begin{cases} \dot{Y}^1(t) = f(t)x^1(t) + g(t), \\ Y^1(0) = 0. \end{cases} \tag{1.31}$$

Here the coefficients $a(\cdot)$, $b(\cdot)$, $\tilde{b}(\cdot)$, $c(\cdot)$, $\tilde{c}(\cdot)$, $f(\cdot)$, $g(\cdot)$, $h(\cdot)$, $h^{-1}(\cdot)$, $A(\cdot)$, $B(\cdot)$, $C(\cdot)$, $\tilde{C}(\cdot)$, $D(\cdot)$, and $\tilde{D}(\cdot)$ are uniformly bounded deterministic functions; x_0 and F are constants; and $\xi \in \mathscr{L}^2_{\mathscr{F}_T}(\Omega,\mathbb{R})$.

According to Theorem A.1, it is easy to see that (1.28), (1.29), (1.30), and (1.31) admit unique solutions, respectively. If we define

$$\begin{aligned} x^v(t) &= x^0(t) + x^1(t), \quad y^v(t) = y^0(t) + y^1(t), \quad z^v(t) = z^0(t) + z^1(t), \\ \tilde{z}^v(t) &= \tilde{z}^0(t) + \tilde{z}^1(t), \quad Y^v(t) = Y^0(t) + Y^1(t), \end{aligned} \tag{1.32}$$

it follows from Itô's formula and (1.28), (1.29), (1.30), (1.31), and (1.32) that $(x^v(\cdot), y^v(\cdot), z^v(\cdot), \tilde{z}^v(\cdot))$, and $Y^v(\cdot)$ are the unique solutions of

$$\begin{cases} dx^v(t) = \big(a(t)x^v(t) + b(t)v(t) + \tilde{b}(t)\big)\, dt + c(t)dW(t) + \tilde{c}(t)d\tilde{W}(t), \\ -dy^v(t) = \big(A(t)x^v(t) + B(t)y^v(t) + C(t)z^v(t) + \tilde{C}(t)\tilde{z}^v(t) \\ \qquad\qquad + D(t)v(t) + \tilde{D}(t)\big)\, dt - z^v(t)dW(t) - \tilde{z}^v(t)d\tilde{W}(t), \\ x^v(0) = x_0, \quad y^v(T) = Fx^v(T) + G \end{cases} \tag{1.33}$$

and

$$\begin{cases} dY^v(t) = (f(t)x^v(t) + g(t))dt + h(t)dW(t), \\ Y^v(0) = 0, \end{cases} \tag{1.34}$$

where the superscript of $(x^v(\cdot), y^v(\cdot), z^v(\cdot), \tilde{z}^v(\cdot))$, and $Y^v(\cdot)$ emphasizes their dependence on $v(\cdot)$. We say $(x^v(\cdot), y^v(\cdot), z^v(\cdot), \tilde{z}^v(\cdot))$ is the state related to $v(\cdot)$, and $Y^v(\cdot)$ is the corresponding observation.

Let $\mathscr{F}_t^{Y^v} = \sigma\{Y^v(s); 0 \le s \le t\}$ and $\mathscr{F}_t^{Y^0} = \sigma\{Y^0(s); 0 \le s \le t\}$. Then a natural definition of admissible control is $v(\cdot) \in \mathscr{L}^2_{\mathscr{F}^{Y^v}}(0,T;\mathbb{R})$. It implies that we want to determine the control by the observation. However, the circular dependence of the control on the observation leads to an immediate difficulty in determining an optimal control. This is the main reason that (1.33) and (1.34) are split into two parts.

We now give a definition of admissible control to avoid the aforementioned difficulty. Let $\mathscr{U}^0_{ad}$ be the collection of all $\mathscr{F}_t^{Y^0}$-adapted process taking values in $\mathbb{R}$ such that $\mathbb{E}\sup_{0 \le t \le T} v^2(t) < \infty$.

Definition 1.3. A control process $v(\cdot)$ is called admissible, if $v(\cdot) \in \mathscr{U}_{ad}^0$ is $\mathscr{F}_t^{Y^v}$-adapted. The set of all admissible controls is denoted by $\mathscr{U}_{ad}$.

The cost functional is of the form

$$\begin{aligned}J(v(\cdot)) = \frac{1}{2}\mathbb{E}\Big\{\int_0^T & \big[L(t)(x^v(t))^2 + O(t)(y^v(t))^2 + R(t)v^2(t)\\ &+2l(t)x^v(t) + 2o(t)y^v(t) + 2r(t)v(t)\big]\,dt\\ &+M(x^v(T))^2 + 2mx^v(T) + N(y^v(0))^2 + 2ny^v(0)\Big\}.\end{aligned} \tag{1.35}$$

Here the coefficients $L(\cdot) \geq 0$, $O(\cdot) \geq 0$, $R(\cdot) \geq 0$, $l(\cdot)$, $o(\cdot)$, and $r(\cdot)$ are uniformly bounded deterministic functions; and $M \geq 0$, $N \geq 0$, m, and n are constants.

We state the LQ optimal control problem of FBSDE as follows.

Problem (FBLQ). To seek a control $u(\cdot) \in \mathscr{U}_{ad}$ such that

$$J(u(\cdot)) = \min_{v(\cdot)\in\mathscr{U}_{ad}} J(v(\cdot))$$

subject to (1.33) and (1.34). If $u(\cdot)$ is optimal, we use notation $(x(\cdot), y(\cdot), z(\cdot), \tilde{z}(\cdot), Y(\cdot))$ to denote $(x^u(\cdot), y^u(\cdot), z^u(\cdot), \tilde{z}^u(\cdot), Y^u(\cdot))$.

Note that since $\mathscr{U}_{ad}$ depends on $v(\cdot)$, the variational method is not suitable for studying the LQ problem. However, we can prove

$$\min_{v'(\cdot)\in\mathscr{U}_{ad}} J(v'(\cdot)) = \min_{v(\cdot)\in\mathscr{U}_{ad}^0} J(v(\cdot)),$$

based on the fact that $\mathscr{U}_{ad}$ is dense in $\mathscr{U}_{ad}^0$ under the metric of $\mathscr{L}^2_{\mathscr{F}^{Y^0}}(0,T;\mathbb{R})$. Then it suffices to study the optimality of $J(v(\cdot))$ over $\mathscr{U}_{ad}^0$.

Theorem 1.6. *Suppose that $u(\cdot)$ is an optimal control of Problem (FBLQ), in the sense that*

$$\frac{d}{d\varepsilon}J(u(\cdot) + \varepsilon v(\cdot))|_{\varepsilon=0} = 0$$

for any $v(\cdot) \in \mathscr{U}_{ad}$, and $(x(\cdot), y(\cdot), z(\cdot), \tilde{z}(\cdot))$ is the corresponding optimal state. Then the FBSDE

$$\begin{cases} dp(t) = & (B(t)p(t) - O(t)y(t) - o(t))\,dt\\ & \quad +C(t)p(t)dW(t) + \tilde{C}(t)p(t)d\tilde{W}(t),\\ -dq(t) = & (a(t)q(t) - A(t)p(t) + L(t)x(t) + l(t))\,dt\\ & \quad -k(t)dW(t) - \tilde{k}(t)d\tilde{W}(t),\\ p(0) = & -Ny(0) - n, \quad q(T) = -Fp(T) + Mx(T) + m\end{cases}$$

admits a unique solution $(p(\cdot), q(\cdot), k(\cdot), \tilde{k}(\cdot)) \in \mathscr{L}^2_{\mathbb{F}}(0,T;\mathbb{R}^4)$ such that

$$R(t)u(t) - D(t)\mathbb{E}\left[p(t)|\mathscr{F}_t^Y\right] + b(t)\mathbb{E}\left[q(t)|\mathscr{F}_t^Y\right] + r(t) = 0 \tag{1.36}$$

with $\mathscr{F}_t^Y = \sigma\{Y^u(s); 0 \leq s \leq t\}$.

Similar to the proof of Theorem 1.3, we have the following verification theorem.

Theorem 1.7. *Let $u(\cdot) \in \mathscr{U}_{ad}$ satisfy*

$$R(t)u(t) - D(t)\mathbb{E}\left[p(t)\middle|\mathscr{F}_t^Y\right] + b(t)\mathbb{E}\left[q(t)\middle|\mathscr{F}_t^Y\right] + r(t) = 0,$$

where $(x(\cdot), y(\cdot), z(\cdot), \tilde{z}(\cdot), p(\cdot), q(\cdot), k(\cdot), \tilde{k}(\cdot))$ is a solution to

$$\begin{cases} dx(t) = & (a(t)x(t) + b(t)u(t) + \tilde{b}(t))dt + c(t)dW(t) + \tilde{c}(t)d\tilde{W}(t), \\ -dy(t) = & (A(t)x(t) + B(t)y(t) + C(t)z(t) + \tilde{C}(t)\tilde{z}(t) \\ & +D(t)u(t) + \tilde{D}(t))dt - z(t)dW(t) - \tilde{z}(t)d\tilde{W}(t), \\ dp(t) = & (B(t)p(t) - O(t)y(t) - o(t))dt \\ & +C(t)p(t)dW(t) + \tilde{C}(t)p(t)d\tilde{W}(t), \\ -dq(t) = & (a(t)q(t) - A(t)p(t) + L(t)x(t) + l(t))dt \\ & -k(t)dW(t) - \tilde{k}(t)d\tilde{W}(t), \\ x(0) = & x_0, \quad y(T) = Fx(T) + G, \\ p(0) = & -Ny(0) - n, \quad q(T) = -Fp(T) + Mx(T) + m \end{cases}$$

with (1.36). Then $u(\cdot)$ is an optimal control of Problem (FBLQ).

Furthermore, we prove the uniqueness of optimal control by the parallelogram law.

Theorem 1.8. *Let $R(\cdot) > 0$ and $R^{-1}(\cdot)$ be uniformly bounded, deterministic functions. If $u(\cdot)$ is an optimal control of Problem (FBLQ), then $u(\cdot)$ is unique.*

The second example in Chapter 5 is an LQ model of Case A2.

Problem (BLQ). To minimize the cost functional

$$J(v(\cdot)) = \frac{1}{2}\mathbb{E}\left\{\int_0^T \left[O(t)(y^v(t))^2 + R(t)v^2(t)\right]dt + N(y^v(0))^2 + 2ny^v(0)\right\}$$

over $\mathscr{U}_{ad}$ subject to the state equation

$$\begin{cases} -dy^v(t) = & \left(B(t)y^v(t) + C(t)z^v(t) + \tilde{C}(t)\tilde{z}^v(t) + D(t)v(t)\right)dt \\ & -z^v(t)dW(t) - \tilde{z}^v(t)d\tilde{W}(t), \\ y^v(T) = & \xi. \end{cases}$$

Here $O(\cdot) \geq 0$, $R(\cdot) \geq 0$, $B(\cdot)$, $C(\cdot)$, $\tilde{C}(\cdot)$, $D(\cdot)$, and $R^{-1}(\cdot)$ are uniformly bounded deterministic functions; $N \geq 0$ and n are constants; ξ is a square-integrable random variable; and $(W(\cdot), \tilde{W}(\cdot))$ is a 2-dimensional standard Brownian motion. Suppose that $W(t)$ is observable at time t.

The observation process here looks simple, but the traditional separation principle still does not work. Fortunately, the backward separation approach introduced in this book can be used to solve this example explicitly. We first introduce two ordinary differential equations (ODEs)

$$\begin{cases} \dot{\alpha}(t) - \left(2B(t) + C^2(t) + \tilde{C}^2(t)\right)\alpha(t) - \dfrac{1}{R(t)}D^2(t)\alpha^2(t) + O(t) = 0, \\ \alpha(0) = -N, \end{cases} \tag{1.37}$$

and

$$\begin{cases} \dot{\beta}(t) - \left(B(t) + C^2(t) + \tilde{C}^2(t) + \frac{1}{R(t)} D^2(t)\alpha(t) \right) \beta(t) = 0, \\ \beta(0) = -n. \end{cases} \tag{1.38}$$

We assume that the solution $\alpha(\cdot)$ of (1.37) satisfies

$$\frac{1}{\alpha(t)}\tilde{C}^2(t) + \frac{1}{R(t)}D^2(t) \geq 0.$$

Combining the theory of FBSDE with Theorems 1.1 and 1.2, we have

Theorem 1.9. *The optimal control of Problem (BLQ) is given uniquely by*

$$u(t) = \frac{1}{R(t)} D(t)(\alpha(t)\hat{y}(t) + \beta(t)),$$

where $\alpha(\cdot)$, $\beta(\cdot)$ *and* $(\hat{p}(\cdot), \hat{y}(\cdot), \hat{z}(\cdot))$ *are the solutions of (1.37), (1.38), and*

$$\begin{cases} d\hat{p}(t) = (B(t)\hat{p}(t) - O(t)\hat{y}(t))\,dt + C(t)\hat{p}(t)dW(t), \\ -d\hat{y}(t) = \left(B(t)\hat{y}(t) + C(t)\hat{z}(t) + \tilde{C}(t)\hat{\tilde{z}}(t) + D(t)u(t)\right)dt - \hat{z}(t)dW(t), \\ \hat{p}(0) = -Ny(0) - n, \quad \hat{y}(T) = \hat{\xi}. \end{cases}$$

By applying Theorem 1.1 to the BSDEs in Theorems 1.4, 1.6, and 1.7, optimal feedback controls of these problems can be obtained in particular cases. However, it is hard to get analytical solutions in general cases. Therefore, it is highly desirable to establish some numerical schemes for these problems.

1.4 Notes

The operation of solving a standard LQ Gaussian control model with incomplete information is separated into two stages in sequence: to compute the filtering of the state first, and then to solve a complete information LQ control problem driven by the filtering. The optimal control is designed as the linear feedback of the filtering of the optimal state. One feature of the optimal control is that the coefficients of the feedback are the same as those of the LQ Gaussian control model with complete information. This feature is known as the certainty equivalence principle, originally introduced in the economic paper of Simon [76]. Note that the certainty equivalence principle does not hold for general cost functionals.

In 1958, Kalman and Koepcke [39] raised a question of "whether the separate optimization of statistical prediction and control system performance yields a system which is optimal in the over-all sense." Joseph and Tou [33] answered the question by a discrete time combined problem of estimate and LQ Gaussian control. In 1968, Wonham [96] proved a more general separation result in continuous time, i.e., the combined problem of control and filtering, for a Gaussian dynamic system observed via a noisy linear system with more general cost functional, can still be separated into two independent problems of filtering and control, respectively. This result is improved by many scholars under various setups and is referred to as the (traditional) separation principle, which is one of the primary results in the theory of stochastic control. See, e.g., Davis [15] and Menaldi [55] for more information.

For a non-Gaussian dynamic control system with partially observable information, however, the separation principle does not hold usually. We may then reduce the control problem to the one with completely observable information of the Zakai's equation. The problem can be regarded as an optimization problem driven by a stochastic partial differential equation (SPDE) in infinite dimensional space, which is essentially difficult to study. See, e.g., the monograph of Bensoussan [6] for a systematic account on the problem.

In 2008, Wang and Wu [84] discovered originally an approach of decoupling control and filtering when they solved a partially observable LQ optimal control problem driven by SDE. The approach allows us to study a combined problem of control and filtering by first deducing the optimal control of an augmented dimensional optimality problem and then computing the optimal filtering of the corresponding Hamiltonian system. Since in the current approach, the filtering and the control are decoupled in the opposite order of the traditional separation principle, [84] called it the backward separation approach. The approach is applicable to a board class of nonlinear control systems and can be used to solve them in a more effective way. There are some references on the topic. We mention in particular Wang et al. [88, 89, 95].

Chapter 2
Filtering of BSDE and FBSDE

In this chapter, we develop some filtering results for the solutions to BSDEs and FBSDEs, which play an important role in studying the optimal control with incomplete information. We first state a theorem on the stochastic filtering of a general stochastic process. The proof of that result can be found in Liptser and Shiyayev [49], so we omit it here. Then, we apply this result to the stochastic filtering for the solutions to BSDEs in Section 3.2 and to those for FBSDEs in Section 3.3.

2.1 Stochastic Filtering of Stochastic Processes

Consider a stochastic process

$$x(t) = x(0) + \int_0^t b(s)ds + m(t), \tag{2.1}$$

where $m(\cdot)$ is an $\mathscr{F}_t$-martingale, and $b(\cdot)$ is a stochastic process with

$$\mathbb{P}\left(\int_0^T |b(s)|ds < \infty\right) = 1.$$

Assume that $x(\cdot)$ is observed via an Itô process

$$Y(t) = Y(0) + \int_0^t h(s)ds + \int_0^t f(s,Y)dW(s).$$

Here $W(\cdot)$ is a 1-dimensional standard Brownian motion defined on $(\Omega, \mathscr{F}, (\mathscr{F}_t), P)$; $h:\ \Omega \times [0,T] \to \mathbb{R}$ and $f:\ [0,T] \times \mathbb{R} \to \mathbb{R}$ satisfy

$$\mathbb{P}\left(\int_0^T |h(s)|ds < \infty\right) = 1, \quad \mathbb{P}\left(\int_0^T |f(s,Y)|^2 ds < \infty\right) = 1$$

G. Wang et al., *An Introduction to Optimal Control of FBSDE with Incomplete Information*, SpringerBriefs in Mathematics,
https://doi.org/10.1007/978-3-319-79039-8_2

with $f(t,Y)$, $Y \in C([0,T];\mathbb{R})$, being $\mathscr{B}_t$-measurable for each $0 \le t \le T$. Furthermore, we assume that for any Y, Y_1, $Y_2 \in C([0,T];\mathbb{R})$, $0 \le t \le T$, there are three constants C, C_1, C_2 and a nondecreasing right continuous function $0 \le K(t) \le 1$ such that

$$|f(t,Y)|^2 \le C_1 \int_0^t |1+Y(s)|^2 dK(s) + C_2|1+Y(t)|^2$$

and

$$|f(t,Y_1)-f(t,Y_2)|^2 \le C_1 \int_0^t |Y_1(s)-Y_2(s)|^2 dK(s) + C_2|Y_1(t)-Y_2(t)|^2.$$

For a stochastic process $X(t)$, we call

$$\hat{X}(t) \equiv \mathbb{E}\left[X(t)|\mathscr{F}_t^Y\right]$$

the optimal filtering of $X(t)$ based on $Y(\cdot)$ up to time t, where $\mathscr{F}_t^Y = \sigma\{Y(s); 0 \le s \le t\}$. We now state the filtering equation of $x(t)$ given in (2.1), whose proof can be found in Liptser and Shiyayev (Theorem 8.1 of [49]).

Theorem 2.1. *Let*

$$\sup_{0 \le t \le T} \mathbb{E}x^2(t) < \infty, \quad \mathbb{E}\int_0^T [b^2(t)+h^2(t)]dt < \infty, \quad f^2(t,Y) \ge C > 0.$$

Then the optimal filtering of $x(t)$ satisfies

$$\hat{x}(t) = \hat{x}(0) + \int_0^t \hat{b}(s)ds + \int_0^t \left[\hat{D}(s) + \frac{\widehat{(xh)}(s) - \hat{x}(s)\hat{h}(s)}{f(s,Y)}\right] d\hat{W}(s),$$

where $(xh)(t) = x(t)h(t)$,

$$\hat{W}(t) = \int_0^t \frac{dY(s) - \hat{h}(s)ds}{f(s,Y)}$$

is a Brownian motion, and $D(t)$ is the stochastic process given by

$$D(t) = \frac{d\langle m, W\rangle_t}{dt}.$$

2.2 Stochastic Filtering for BSDE

Suppose that the stochastic process $(y(\cdot), z_1(\cdot), z_2(\cdot))$ is governed by a BSDE

$$\begin{aligned} y(t) = \xi &+ \int_t^T g(s,y(s),z_1(s),z_2(s))ds \\ &- \int_t^T z_1(s)dW_1(s) - \int_t^T z_2(s)dW_2(s), \end{aligned} \tag{2.2}$$

where $(W_1(\cdot), W_2(\cdot))$ is a 2-dimensional standard Brownian motion, $\xi \in \mathscr{L}^2_{\mathbb{F}}(\Omega;\mathbb{R})$, and $g : [0,T] \times \Omega \times \mathbb{R}^3 \to \mathbb{R}$ is a given function. Equation (2.2) admits a unique solution $(y(\cdot), z_1(\cdot), z_2(\cdot)) \in \mathscr{L}^2_{\mathbb{F}}(0,T;\mathbb{R}^3)$ under Conditions (Ha.1–Ha.2). Note that the solution involves $(z_1(\cdot), z_2(\cdot))$, which can be regarded as a control term to the equation such that the adapted solutions exist. Next, we assume that the observable process $Y(\cdot)$ is the Itô process given by

$$Y(t) = \int_0^t h(s)ds + \int_0^t f(s)dW_1(s), \tag{2.3}$$

where $f : [0,T] \to \mathbb{R}$ and $h : \Omega \times [0,T] \to \mathbb{R}$ are measurable mappings. The optimal nonlinear filtering is to compute $\hat{X}(t) = \mathbb{E}[X(t)|\mathscr{F}_t^Y]$, where $X = y$, z_1 and z_2. Since $z_i(\cdot)$ can be calculated by the Malliavin derivatives of $y(\cdot)$ with respect to $W_i(\cdot)$ $(i = 1,2)$ (see, e.g., El Karoui et al. [19]), we focus on the filtering of $y(\cdot)$. For any $t \in [0,T]$, we adopt the following notations for simplification of the presentation

$$\begin{aligned}
\hat{h}(t) &= \mathbb{E}[h(t)|\mathscr{F}_t^Y],\\
\hat{g}(t) &= \mathbb{E}[g(t,y(t),z_1(t),z_2(t))|\mathscr{F}_t^Y],\\
\widehat{(yh)}(t) &= \mathbb{E}[y(t)h(t)|\mathscr{F}_t^Y].
\end{aligned}$$

Also, we need the following assumption on the coefficients of (2.2) and (2.3).

(H2.1) *The function $g(\cdot,0,0,0) \in \mathscr{L}^2_{\mathscr{F}}(0,T;\mathbb{R})$, and g is Lipschitz in (y,z_1,z_2) uniformly for $(\omega,t) \in \Omega \times [0,T]$. f is bounded and deterministic, and f^{-1} is also bounded. h is in $\mathscr{L}^2_{\mathbb{F}}(0,T;\mathbb{R})$.*

We now state the main result of this section, which plays an important role in the study of incomplete information stochastic control for BSDE.

Theorem 2.2. *Under (H2.1), the optimal filtering $\hat{y}(\cdot)$ is governed by*

$$\begin{aligned}
\hat{y}(t) &= \mathbb{E}[\xi|\mathscr{F}_T^Y] + \int_t^T \hat{g}(s)ds\\
&\quad - \int_t^T \left\{ \hat{z}_1(s) + \frac{1}{f(s)}\left[\widehat{(yh)}(s) - \hat{y}(s)\hat{h}(s)\right]\right\} d\hat{W}(s),
\end{aligned} \tag{2.4}$$

where

$$\hat{W}(t) = \int_0^t \frac{1}{f(s)}\left(dY(s) - \hat{h}(s)ds\right) \tag{2.5}$$

is a 1-dimensional standard Brownian motion defined on $(\Omega, \mathscr{F}^Y, (\mathscr{F}_t^Y)_{0\le t\le T}, \mathbb{P})$.

Proof. Equation (2.2) admits a unique solution $(y(\cdot), z_1(\cdot), z_2(\cdot)) \in \mathscr{L}^2_{\mathbb{F}}(0,T;\mathbb{R}^3)$, so $y(\cdot)$ can be rewritten as an Itô process as follows:

$$\begin{aligned}
y(t) &= y(0) - \int_0^t g(s,y(s),z_1(s),z_2(s))ds\\
&\quad + \int_0^t z_1(s)dW_1(s) + \int_0^t z_2(s)dW_2(s).
\end{aligned} \tag{2.6}$$

Now (2.6) and (2.3) can be regarded as the state equation and the observation equation, respectively. Using Theorem 2.1, we have

$$\hat{y}(t) = \hat{y}(0) - \int_0^t \hat{g}(s)ds + \int_0^t \left\{ \hat{z}_1(s) + \frac{1}{f(s)} \left[\widehat{(yh)}(s) - \hat{y}(s)\hat{h}(s) \right] \right\} d\hat{W}(s), \quad (2.7)$$

where $\hat{W}(\cdot)$ is given by (2.5). Similarly, we have

$$\hat{y}(T) = \hat{y}(0) - \int_0^T \hat{g}(s)ds + \int_0^T \left\{ \hat{z}_1(s) + \frac{1}{f(s)} \left[\widehat{(yh)}(s) - \hat{y}(s)\hat{h}(s) \right] \right\} d\hat{W}(s). \quad (2.8)$$

Subtracting (2.8) from (2.7), we obtain that

$$\hat{y}(t) = \hat{y}(T) + \int_t^T \hat{g}(s)ds - \int_t^T \left\{ \hat{z}_1(s) + \frac{1}{f(s)} \left[\widehat{(yh)}(s) - \hat{y}(s)\hat{h}(s) \right] \right\} d\hat{W}(s).$$

The verification of the terminal condition $\hat{y}(T) = \mathbb{E}[\xi | \mathscr{F}_T^Y]$ is trivial. Thus the proof is completed. □

The following result is an immediate consequence of Theorem 2.2.

Corollary 2.1. *Under (H2.1), if $g_1 : [0,T] \times \mathbb{R}^2 \to \mathbb{R}$ and $h : \Omega \times [0,T] \to \mathbb{R}$ satisfy*

$$\hat{g}(t) = g_1(t, \hat{y}(t), \hat{z}_1(t)) \text{ and } \widehat{(yh)}(t) = \hat{y}(t)\hat{h}(t),$$

respectively, then the optimal filtering $(\hat{y}(\cdot), \hat{z}_1(\cdot))$ is a solution of the following backward filtering equation:

$$\hat{y}(t) = \mathbb{E}[\xi | \mathscr{F}_T^Y] + \int_t^T g_1(s, \hat{y}(s), \hat{z}_1(s))ds - \int_t^T \hat{z}_1(s)d\hat{W}(s),$$

where

$$\hat{W}(t) = \int_0^t \frac{1}{f(s)} \left(dY(s) - \hat{h}(s)ds \right)$$

is a 1-dimensional standard Brownian motion defined on $(\Omega, \mathscr{F}^Y, (\mathscr{F}_t^Y)_{0 \le t \le T}, \mathbb{P})$.

In what follows, suppose that the conditional probability distribution of $y(t)$ based on $\mathscr{F}_t^Y$ has the density

$$\psi(t,x) = \frac{d\mathbb{P}(y(t) \le x | \mathscr{F}_t^Y)}{dx}, \quad (t,x) \in [0,T] \times \mathbb{R},$$

which is measurable in (t,x,ω). We proceed to deriving the equation satisfied by this conditional density. Note that the nonlinear filtering $\hat{y}(\cdot)$ can be represented by

$$\hat{y}(t) = \mathbb{E}[y(t) | \mathscr{F}_t^Y] = \int_{-\infty}^{\infty} x\psi(t,x)dx.$$

Next, we assume that the observation function $h(s)$ in (2.3) depends on the signal in a deterministic way, namely, we abuse notation a bit, there is a function $h: [0,T] \times \mathbb{R} \to \mathbb{R}$ such that observation equation (2.3) is replaced by

$$Y(t) = \int_0^t h(s, y(s))ds + \int_0^t f(s)dW_1(s). \tag{2.9}$$

We introduce the following assumptions:

(H2.2) *The function (to be used in Theorem 2.3 below as test function for filtering) $\varphi(\cdot) : \mathbb{R} \to \mathbb{R}$ and its derivatives up to order 2 are uniformly bounded.*

(H2.3) *The function $\varphi(\cdot) : \mathbb{R} \to \mathbb{R}$ has compact support.*

(H2.4) *The functions $g : [0,T] \times \mathbb{R}^3 \to \mathbb{R}$ and $h : [0,T] \times \mathbb{R} \to \mathbb{R}$ are Borel measurable, and h is Lipschitz in $x \in \mathbb{R}$ uniformly for $t \in [0,T]$.*

(H2.5) *The solution of (2.2) satisfies $z_i(t) = g_i(t, y(t))$, where $g_i : [0,T] \times \mathbb{R} \to \mathbb{R}$ $(i = 1,2)$ are also Borel measurable.*

(H2.6) *The partial derivatives*

$$[\psi(t,x)g(t,x,y,z)]_x, \qquad [yh(t,x)]_x,$$

$$[\psi(t,x)(h(t,x) - \hat{h}(t,x))g(t,x,y,z)]_x,$$

$$[(y^2 + z^2)\psi(t,x)]_{xx} \text{ and } [(y^2 + z^2)(h(t,x) - \hat{h}(t,x))\psi(t,x)]_{xx}$$

exist.

(H2.7)

$$\int_0^T \int_{-\infty}^{\infty} |\varphi(x)\mathscr{L}^*\psi(t,x)|dxdt < \infty,$$

$$\mathbb{E}\int_0^T \int_{-\infty}^{\infty} \varphi^2(x) \left[\psi(t,x)\left(h(t,x) - \hat{h}(t,x)\right) + \mathscr{N}^*\psi(t,x)\right]^2 dxdt < \infty$$

with the notations

$$\mathscr{L}^*\psi(t,x) = -[\psi(t,x)g(t,x,y,z)]_x - \frac{1}{2}[\psi(t,x)(y^2 + z^2)]_{xx}$$

and

$$\mathscr{N}^*\psi(t,x) = -[\psi(t,x)y]_x.$$

Note that (H2.2), (H2.3), (H2.4), (H2.6), and (H2.7) are standard in the theory of nonlinear filtering, while (H2.5) is reasonable under some constraints on ξ and g. We give an example below for which (H2.5) holds.

Example 2.1. We consider the BSDE

$$\begin{aligned} y(t) = \xi + &\int_t^T (a(s)y(s) + b_1(s)z_1(s) + b_2(s)z_2(s))ds \\ &- \int_t^T z_1(s)dW_1(s) - \int_t^T z_2(s)dW_2(s), \end{aligned}$$

where $a(\cdot)$, $b_i(\cdot)$ $(i=1,2)$ are bounded and deterministic, and

$$\xi = \exp\left(\sum_{i=1}^{2}\int_0^T f_i(t)dW_i(t)\right).$$

Then, (H2.5) holds.

Proof. It is easy to see that the BSDE has a unique solution $(y(\cdot),z_1(\cdot),z_2(\cdot)) \in \mathscr{L}^2_{\mathbb{F}}(0,T;\mathbb{R}^3)$. In fact, the solution can be represented as

$$y(t) = \mathbb{E}[\xi x(T)|\mathscr{F}_t]$$

with

$$\begin{aligned} x(T) = \exp\Bigg\{&\int_0^T \left[a(t) - \frac{1}{2}\left(b_1^2(t)+b_2^2(t)\right)\right]dt \\ &+\int_0^T b_1(t)dW_1(t) + \int_0^T b_2(t)dW_2(t)\Bigg\}. \end{aligned}$$

According to Proposition A.1, $z_i(\cdot)$ is expressed by

$$\begin{aligned} z_i(t) &= D_t^{(W_i)}\mathbb{E}[\xi x(T)|\mathscr{F}_t] \\ &= \mathbb{E}\left[x(T)D_t^{(W_i)}\xi|\mathscr{F}_t\right] + b_i(t)y(t), \end{aligned}$$

where $D_t^{(W_i)}\eta$ stands for the Malliavin derivative of η with respect to $W_i(\cdot)$ $(i=1,2)$. Note that $D_t^{(W_i)}\xi = \xi f_i(t)$. Thus, $z_i(t) = (f_i(t)+b_i(t))y(t)$, and hence (H2.5) holds. □

Theorem 2.3. *(i) If (H2.1), (H2.2), and (H2.5) hold, then the optimal filtering $\hat{\varphi}(y(t))$ satisfies*

$$\begin{aligned} \hat{\varphi}(y(t)) = \mathbb{E}[\varphi(\xi)|\mathscr{F}_T^Y] &+ \int_t^T \widehat{\mathscr{L}\varphi}(y(s))ds \\ &- \int_t^T \left\{\widehat{\mathscr{N}\varphi}(y(s)) + \frac{1}{f(s)}\left[\widehat{\varphi h}(s,y(s)) - \hat{\varphi}(y(s))\hat{h}(s,y(s))\right]\right\}d\hat{W}(s), \end{aligned} \tag{2.10}$$

where

$$\begin{aligned} \widehat{\mathscr{L}\varphi}(y(s)) &= \mathbb{E}[\mathscr{L}\varphi(y(s))|\mathscr{F}_s^Y], \\ \widehat{\mathscr{N}\varphi}(y(s)) &= \mathbb{E}[\mathscr{N}\varphi(y(s))|\mathscr{F}_s^Y], \\ \widehat{\varphi h}(s,y(s)) &= \mathbb{E}[\varphi(y(s))h(s,y(s))|\mathscr{F}_s^Y], \end{aligned}$$

with

$$\begin{aligned} \mathscr{L}\varphi(y(s)) &= \varphi_x(y(s))g(s,y(s),z_1(s),z_2(s)) - \frac{1}{2}\varphi_{xx}(y(s))(z_1^2(s)+z_2^2(s)), \\ \mathscr{N}\varphi(y(s)) &= \varphi_x(y(s))z_1(s), \end{aligned}$$

and

$$\hat{W}(s) = \int_0^s \frac{1}{f(t)} \left(dY(t) - \hat{h}(t, y(t))\right) dt.$$

(ii) If (H2.1), (H2.3), (H2.4), (H2.6), (H2.7), (H2.8), (H2.9), and (H2.10) hold, then the conditional density $\psi(t,x)$ *satisfies*

$$\begin{aligned}\psi(t,x) = \psi(T,x) + \int_t^T \mathscr{L}^* \psi(s,x)ds - \int_t^T \Big[\mathscr{N}^* \psi(s,x) \\ + \frac{1}{f(s)}\psi(s,x)\left(h(s,x) - \int_{-\infty}^{\infty} h(s,x)\psi(s,x)dx\right)\Big] d\hat{W}(s).\end{aligned} \tag{2.11}$$

Proof. (i) Applying Itô's formula to $\varphi(y(t))$, we have

$$\begin{aligned}\varphi(y(t)) =& \varphi(\xi) + \int_t^T \mathscr{L}\varphi(y(s))ds \\ &- \int_t^T \varphi_x(y(s))z_1(s)dW_1(s) - \int_t^T \varphi_x(y(s))z_2(s)dW_2(s).\end{aligned}$$

By the uniqueness of $(y(\cdot), z_1(\cdot), z_2(\cdot))$, Theorems 2.1 and 2.2, the nonlinear filtering equation (2.10) is obtained directly.

(ii) Due to (H2.5), (2.10) can be rewritten as

$$\int_{-\infty}^{\infty} \varphi(x)\psi(t,x)dx = \int_{-\infty}^{\infty} \varphi(x)\psi(T,x)dx + I - II, \tag{2.12}$$

where

$$\begin{aligned}I &= \int_t^T \int_{-\infty}^{\infty} \mathscr{L}\varphi(x)\psi(s,x)dxds, \\ II &= \int_t^T \int_{-\infty}^{\infty} \left[\mathscr{N}\varphi(x) + \frac{1}{f(s)}\varphi(x)\left(h(s,x) - \hat{h}(s,x)\right)\right] \psi(s,x)dxd\hat{W}(s).\end{aligned}$$

It follows from integration by parts and Fubini's theorem that

$$I = \int_{-\infty}^{\infty} \int_t^T \varphi(x)\mathscr{L}^* \psi(s,x)dsdx.$$

Similarly,

$$II = \int_{-\infty}^{\infty} \int_t^T \varphi(x)\left[\mathscr{N}^* \psi(s,x) + \frac{1}{f(s)}\psi(s,x)\left(h(s,x) - \hat{h}(s,x)\right)\right] d\hat{W}(s)dx.$$

Substituting the above two identities into (2.12), we obtain

$$\begin{aligned}
&\int_{-\infty}^{\infty} \varphi(x)\psi(t,x)dx \\
&= \int_{-\infty}^{\infty} \varphi(x)\left\{\psi(T,x) + \int_t^T \mathscr{L}^*\psi(s,x)ds \right. \\
&\quad \left. - \int_t^T \left[\mathscr{N}^*\psi(s,x) + \frac{1}{f(s)}\psi(s,x)\left(h(s,x)-\hat{h}(s,x)\right)\right] d\hat{W}(s)\right\} dx.
\end{aligned}$$

The arbitrariness of $\varphi(\cdot)$ implies (2.11). The proof is then completed. □

We emphasize that (2.11) is a new kind of backward stochastic partial differential equation (BSPDE). Since the noise term is very complicated, it is not easy to prove the existence and uniqueness of solution to the equation. We pose it here as an open problem.

2.3 Stochastic Filtering for FBSDE

In this section, we first introduce the four-step scheme in solving FBSDEs. As an example to this scheme, we then consider an optimization problem. To obtain an explicit solution, we will apply Girsanov's transformation to convert it to an LQ control problem which can be solved in terms of an FBSDE without control variable. Finally, we study the stochastic filtering problem for this FBSDE based on a linear observation equation.

Consider a fully coupled FBSDE

$$\begin{cases}
dx(t) = b(t,x(t),y(t),z(t))dt + \sigma(t,x(t),y(t),z(t))dW(t), \\
-dy(t) = g(t,x(t),y(t),z(t))dt - z(t)dW(t), \\
x(0) = x_0, \quad y(T) = f(x(T)),
\end{cases} \tag{2.13}$$

where $b,g:[0,T]\times\mathbb{R}^{n+n+n\times m}\to\mathbb{R}^n$, $\sigma:[0,T]\times\mathbb{R}^{n+n+n\times m}\to\mathbb{R}^{n\times m}$, $f:\Omega\times\mathbb{R}^n\to\mathbb{R}^n$ are continuous functions; $W(\cdot)$ is an m-dimensional standard Brownian motion defined on the filtered probability space $(\Omega,\mathscr{F},(\mathscr{F}_t)_{0\le t\le T},\mathbb{P})$; $\mathscr{F}_t$ is the natural filtration generated by $W(\cdot)$, and $x_0\in\mathbb{R}^n$. Under Conditions (Ha.3-Ha.4), there is a unique solution $(x(\cdot),y(\cdot),z(\cdot))$ to (2.13). Furthermore, using the four-step scheme (see, e.g., Yong and Zhou [109]), $y(\cdot)$ and $z(\cdot)$ can be expressed as a functional of $x(\cdot)$, respectively. Indeed,

$$\begin{aligned}
y(t) &= U(t,x(t)), \\
z(t) &= Z(t,x(t),U(t,x(t)),U_x(t,x(t))),
\end{aligned}$$

where U, Z, and $x(\cdot)$ satisfy

$$\begin{cases} U_t^\ell + \frac{1}{2} tr\left[U_{xx}^\ell(\sigma\sigma^\top)(t,X,U,Z(t,X,U,U_x))\right] + \langle b(t,X,U,U_x),U_X^\ell\rangle \\ \quad + g^\ell(t,X,U,U_x) = 0, \quad (t,X) \in (0,T)\times\mathbb{R}^n, \quad \ell = 1,\cdots,n, \\ U(T,X) = f(X), \quad X \in \mathbb{R}^n, \end{cases}$$

$$Z(t,X,\bar{X},\tilde{X}) = \tilde{X}\sigma(t,X,\bar{X},Z(t,X,\bar{X},\tilde{X})), \quad (t,X,\bar{X},\tilde{X}) \in [0,T]\times\mathbb{R}^{n+n+n\times n},$$

and

$$\begin{cases} dx(t) = \tilde{b}(t,x(t))dt + \tilde{\sigma}(t,x(t))dW(t), \\ x(0) = x_0 \end{cases}$$

with

$$\tilde{b}(t,X) = b(t,X,U(t,X),Z(t,X,U(t,X),U_x(t,X))),$$
$$\tilde{\sigma}(t,X) = \sigma(t,X,U(t,X),Z(t,X,U(t,X),U_x(t,X))).$$

According to the relationship between $(y(\cdot),z(\cdot))$, and $x(\cdot)$, it suffices to compute the optimal filtering of $x(\cdot)$. The detailed arguments are omitted due to the page limit.

To elaborate the above analysis, we present a simple example on filtering of stochastic Hamiltonian system arising from a stochastic control problem. Specifically, let us consider a 1-dimensional control system, whose evolution is described by

$$\begin{cases} dx(t) = (A(t)x(t) + B(t)v(t))dt + C_1(t)d\tilde{W}_1(t) + C_2(t)d\tilde{W}_2(t), \\ x(0) = x_0, \end{cases} \tag{2.14}$$

where $v(\cdot)$ is an element of the set

$$\mathscr{U}_{ad} = \left\{ v(\cdot) \middle| v(t) \text{ is an } \mathscr{F}_t\text{-adapted process valued in } \mathbb{R} \right.$$
$$\left. \text{and satisfies } \mathbb{E}\int_0^T v^4(t)dt < \infty \right\}.$$

Suppose that the cost functional is given by

$$J(v(\cdot)) = \bar{y}(0),$$

where $\bar{y}(\cdot)$ is a solution to the BSDE

$$\begin{cases} -d\bar{y}(t) = (a(t)x^2(t) + b(t)\bar{y}(t) + f_1(t)\bar{z}_1(t) + f_2(t)\bar{z}_2(t) + c(t)v^2(t))dt \\ \qquad - \bar{z}_1(t)d\tilde{W}_1(t) - \bar{z}_2(t)d\tilde{W}_2(t), \\ \bar{y}(T) = x^2(T). \end{cases} \tag{2.15}$$

Here $a(\cdot) \geq 0$, $c(\cdot) \geq \varepsilon > 0$, $A(\cdot)$, $B(\cdot)$, $C_1(\cdot)$, $C_2(\cdot)$, $f_1(\cdot)$, and $f_2(\cdot)$ are uniformly bounded, deterministic functions. For any $v(\cdot) \in \mathscr{U}_{ad}$, it is easy to see that

$$\mathbb{E}\tilde{y}^2(T) < \infty,$$

and thus, there exists a unique solution to (2.14) and (2.15), respectively. Since the drift term in (2.15) contains $(\bar{z}_1(\cdot), \bar{z}_2(\cdot))$, it causes us some trouble to express the cost functional $J(v(\cdot))$. To simplify it, we define a probability measure Q on the space $(\Omega, \mathscr{F})$ by

$$\frac{dQ}{dP} = \exp\left\{\int_0^T f_1(t)d\tilde{W}_1(t) + \int_0^T f_2(t)d\tilde{W}_2(t) - \frac{1}{2}\int_0^T (f_1^2(t) + f_2^2(t))dt\right\}.$$

It follows from Girsanov's theorem that $(W_1(\cdot), W_2(\cdot))$ defined by

$$W_1(t) = \tilde{W}_1(t) - \int_0^t f_1(s)ds \quad \text{and} \quad W_2(t) = \tilde{W}_2(t) - \int_0^t f_2(s)ds$$

is a 2-dimensional Brownian motion defined on the stochastic basis $(\Omega, \mathscr{F}, (\mathscr{F}_t), Q)$. Then we can rewrite (2.14) and (2.15) as

$$\begin{cases} dx(t) = (A(t)x(t) + B(t)v(t) + C_1(t)f_1(t) + C_2(t)f_2(t))dt \\ \qquad\qquad + C_1(t)dW_1(t) + C_2(t)dW_2(t), \\ x(0) = x_0, \end{cases} \tag{2.16}$$

$$\begin{cases} -d\bar{y}(t) = (a(t)x^2(t) + b(t)\bar{y}(t) + c(t)v^2(t))dt \\ \qquad\qquad - \bar{z}_1(t)dW_1(t) - \bar{z}_2(t)dW_2(t), \\ \bar{y}(T) = x^2(T). \end{cases} \tag{2.17}$$

Integrating on both sides of (2.17), we get

$$J(v(\cdot)) = \bar{y}(0) = \mathbb{E}_Q\left[\int_0^T e^{\int_0^t b(s)ds}(a(t)x^2(t) + c(t)v^2(t))dt + e^{\int_0^T b(t)dt}x^2(T)\right].$$

Then minimizing the cost functional subject to $v(\cdot) \in \mathscr{U}_{ad}$ and (2.16) formulates a complete information LQ optimal control problem. Since the drift term in (2.16) contains the deterministic function $C_1(\cdot)f_1(\cdot) + C_2(\cdot)f_2(\cdot)$, the classical technique of completing squares cannot be used directly to solve the control problem. However, stochastic maximum principle (see, e.g., Chapters 3–5) provides an alternative tool. According to the maximum principle, we derive the desired optimal control

$$u(t) = -\frac{1}{2}B(t)c^{-1}(t)e^{-\int_0^t b(s)ds}y(t),$$

where the adjoint process $y(\cdot)$ satisfies a Hamiltonian system

$$\begin{cases} dx(t) = \Big(A(t)x(t) - \frac{1}{2}B^2(t)c^{-1}(t)e^{-\int_0^t b(s)ds}y(t) + C_1(t)f_1(t) + C_2(t)f_2(t)\Big)\,dt \\ \qquad\qquad + C_1(t)dW_1(t) + C_2(t)dW_2(t), \\ -dy(t) = \Big(2a(t)e^{\int_0^t b(s)ds}x(t) + A(t)y(t)\Big)\,dt - z_1(t)dW_1(t) - z_2(t)dW_2(t), \\ x(0) = x_0, \quad y(T) = 2e^{\int_0^T b(s)ds}x(T). \end{cases}$$

It follows from Theorem A.3 that there is a unique solution to the above equation. Suppose that $(x(\cdot),y(\cdot),z_1(\cdot),z_2(\cdot))$ cannot be observed directly; however, we can observe a noisy process $Y(\cdot)$ related to $x(\cdot)$, whose dynamic is described by

$$\begin{cases} dY(t) = (D(t)x(t) + F(t)Y(t) + f_2(t)H(t))dt + H(t)dW_2(t), \\ Y(0) = 0, \end{cases} \tag{2.18}$$

where $D(\cdot)$, $F(\cdot)$, $H(\cdot)$, and $H^{-1}(\cdot)$ are uniformly bounded, deterministic functions. Obviously, there exists a unique solution for (2.18).

We now study the filtering $(\hat{x}(t),\hat{y}(t),\hat{z}_1(t),\hat{z}_2(t))$ of $(x(t),y(t),z_1(t),z_2(t))$ with respect to the observation $Y(\cdot)$ up to time t, i.e., we want to derive the explicit expressions for

$$\begin{aligned} \hat{x}(t) &= \mathbb{E}_Q[x(t)|\mathscr{F}_t^Y], \quad \hat{y}(t) = \mathbb{E}_Q[y(t)|\mathscr{F}_t^Y], \\ \hat{z}_1(t) &= \mathbb{E}_Q[z_1(t)|\mathscr{F}_t^Y], \quad \hat{z}_2(t) = \mathbb{E}_Q[z_2(t)|\mathscr{F}_t^Y] \end{aligned} \tag{2.19}$$

and their square error estimates, where

$$\mathscr{F}_t^Y = \sigma\{Y(s); 0 \le s \le t\}.$$

The method used here is first to look for the relationship between $x(\cdot)$ and $(y(\cdot),z(\cdot))$ by the four-step scheme, then to compute $(\hat{x}(\cdot),\hat{y}(\cdot),\hat{z}_1(\cdot),\hat{z}_2(\cdot))$ by traditional filtering theory for SDE.

Set $y(t) = U(t,x(t))$. It follows from the four-step scheme that $z_1(t)$ and $z_2(t)$ can be written as

$$z_1(t) = C_1(t)U_x(t,x(t)), \quad z_2(t) = C_2(t)U_x(t,x(t)), \tag{2.20}$$

where $U(t,x)$ is a classical solution of the PDE

$$\begin{cases} U_t(t,x) + \mathscr{L}U(t,x) + 2a(t)e^{\int_0^t b(s)ds}x + A(t)U(t,x) = 0, \\ U(T,x) = 2e^{\int_0^T b(s)ds}x, \end{cases} \tag{2.21}$$

with

$$\begin{aligned}\mathscr{L}U(t,x) &= \frac{1}{2}(C_1^2(t)+C_2^2(t))U_{xx}(t,x) \\ &+\left(A(t)x-\frac{1}{2}B^2(t)c^{-1}(t)e^{-\int_0^t b(s)ds}U(t,x)\right. \\ &\left.+C_1(t)f_1(t)+C_2(t)f_2(t)\right)U_x(t,x).\end{aligned}$$

Noticing the terminal condition of (2.21), we set

$$U(t,x) = \Pi(t)x+\pi(t),$$

where $\Pi(\cdot)$ and $\pi(\cdot)$ satisfy

$$\begin{cases} \dot{\Pi}(t)+2A(t)\Pi(t)-\frac{1}{2}B^2(t)c^{-1}(t)e^{-\int_0^t b(s)ds}\Pi^2(t)+2a(t)e^{\int_0^t b(s)ds}=0, \\ \Pi(T)=2e^{\int_0^T b(s)ds}, \end{cases} \tag{2.22}$$

and

$$\begin{cases} \dot{\pi}(t)+\left(A(t)-\frac{1}{2}B^2(t)c^{-1}(t)e^{-\int_0^t b(s)ds}\Pi(t)\right)\pi(t) \\ +(C_1(t)f_1(t)+C_2(t)f_2(t))\Pi(t)=0, \\ \pi(T)=0, \end{cases} \tag{2.23}$$

respectively. From the classical ODE theory, we know that there exists a unique solution for (2.22) and (2.23), respectively. Combining (2.20) with (2.23), we get

$$y(t)=\Pi(t)x(t)+\pi(t), \quad z_1(t)=C_1(t)\Pi(t), \quad z_2(t)=C_2(t)\Pi(t), \tag{2.24}$$

where $x(\cdot)$ satisfies

$$\begin{cases} dx(t)= \left[\left(A(t)-\frac{1}{2}B^2(t)c^{-1}(t)\Pi(t)e^{-\int_0^t b(s)ds}\right)x(t)\right. \\ \qquad \left.+C_1(t)f_1(t)+C_2(t)f_2(t)-\frac{1}{2}B^2(t)c^{-1}(t)\pi(t)e^{-\int_0^t b(s)ds}\right]dt \\ \qquad +C_1(t)dW_1(t)+C_2(t)dW_2(t), \\ x(0)=x_0. \end{cases}$$

Obviously,

$$\hat{z}_1(t)=C_1(t)\Pi(t), \quad \hat{z}_2(t)=C_2(t)\Pi(t). \tag{2.25}$$

Then we only need to compute $\hat{x}(t)$ and $\hat{y}(t)$. Let $P(t)=\mathbb{E}_Q(x(t)-\hat{x}(t))^2$ be the square error of the estimate $\hat{x}(t)$. From the fact that $(x(t)-\hat{x}(t)) \perp \mathscr{F}_t^Y$ and $x(t)-\hat{x}(t)$ is Gaussian, we know that $x(t)-\hat{x}(t)$ is independent of $\mathscr{F}_t^Y$. So

$$
\begin{aligned}
P(t) &= \mathbb{E}_Q(x(t)-\hat{x}(t))^2 \\
&= \mathbb{E}_Q[(x(t)-\hat{x}(t))^2|\mathscr{F}_t^Y].
\end{aligned}
$$

Thanks to Theorem 2.1, we obtain

$$
\begin{cases}
d\hat{x}(t) = \Bigg[\left(A(t)-\dfrac{1}{2}B^2(t)c^{-1}(t)\Pi(t)e^{-\int_0^t b(s)ds}\right)\hat{x}(t)+C_1(t)f_1(t) \\
\qquad +C_2(t)f_2(t)-\dfrac{1}{2}B^2(t)c^{-1}(t)\pi(t)e^{-\int_0^t b(s)ds}\Bigg]dt \\
\qquad +(C_2(t)+D(t)H^{-1}(t)P(t))d\bar{W}(t), \\
\hat{x}(0)=x_0,
\end{cases}
\tag{2.26}
$$

$$
\begin{cases}
\dot{P}(t)-2\left(A(t)-\dfrac{1}{2}B^2(t)c^{-1}(t)\Pi(t)e^{-\int_0^t b(s)ds}\right)P(t) \\
+(C_2(t)+D(t)H^{-1}(t)P(t))^2-C_1^2(t)-C_2^2(t)=0, \\
P(0)=0,
\end{cases}
$$

where

$$
\begin{aligned}
\bar{W}(t) &= \int_0^t H^{-1}(s)(dY(s)-D(s)\hat{x}(s)-F(s)Y(s)-f_2(s)H(s))ds \\
&= W_2(t)+\int_0^t D(s)H^{-1}(s)(x(s)-\hat{x}(s))ds
\end{aligned}
$$

is an observable standard Brownian motion defined on $(\Omega,\mathscr{F}^Y,(\mathscr{F}_t^Y),\mathbb{P})$. Furthermore, taking conditional expectations on both sides of (2.24), we get

$$
\hat{y}(t)=\Pi(t)\hat{x}(t)+\pi(t), \tag{2.27}
$$

where $\hat{x}(\cdot)$ is the solution of (2.26). Then we have

Proposition 2.1. *The stochastic filtering process* $(\hat{x}(\cdot),\hat{y}(\cdot),\hat{z}_1(\cdot),\hat{z}_2(\cdot))$ *of the state process* $(x(\cdot),y(\cdot),z_1(\cdot),z_2(\cdot))$ *based on the observation process* $Y(\cdot)$ *is given by (2.26), (2.27), and (2.25).*

2.4 Notes

The filtering problem is to obtain the best linear estimate $\hat{x}(t)$ of an unobservable state $x(t)$ based on the noisy observation data $Y_0^t=\{Y(s);0\le s\le t\}$ related to the state. If $\mathbb{E}x^2(t)<\infty$, then the best estimate $\hat{x}(t)$ of $x(t)$ is equivalent to finding the conditional expectation

$$
\hat{x}(t)=\mathbf{E}[x(t)|\mathscr{F}_t^Y]
$$

with $\mathscr{F}_t^Y = \sigma\{Y(s); 0 \le s \le t\}$. When the estimate depends linearly on the observations, we call it the linear filtering. Otherwise, it is referred to as the nonlinear filtering. In the linear filtering theory, the most celebrated result is the linear quadratic estimation, also known as the Kalman–Bucy filtering. The filtering was discovered and was developed by Rudolf E. Kalman and Richard S. Bucy during the Cold War between North American Treaty Organization and Warsaw Treaty Organization. The Kalman–Bucy filtering works recursively and runs in real time, and thus, it has numerous applications in the fields of aerospace, telecommunication, economics, and so on. As far as we know, the most famous one among these applications is the Apollo Project, where the Kalman–Bucy filtering was used to estimate the trajectories of manned spaceship going to Moon and back.

The filtering equation provided in Theorem 2.1 is one fundamental equation of the nonlinear filtering theory. Lots of known filtering results can be deduced from the equation, say, the Kalman–Bucy filtering. The deduction of the fundamental equation follows the innovation process method, proposed originally by Bode and Shannon [11], whose modern form was presented first by Kailath [34] and Kailath and Frost [35]. Along this line, we get the equation of the conditional probability density $\psi(t,x)$ in Theorem 2.3, which is a new kind of nonlinear backward SPDE. Historically, a similar SPDE was derived early by Stratonovich [77] and Kushner [40] when they studied the condition probability density of a forward SDE. The innovation method achieved its culmination with the famous work of Fujisaki et al. [25]. The filtering equation is called the Kushner–Stratonovich equation or the Kushner–FKK equation. Almost at the same time, Duncan [18], Mortensen [56], and Zakai [112] studied the nonlinear filtering problem by virtue of the Kallianpur–Striebel formula and the unnormalized filtering. They obtained a linear SPDE of the unnormalized filtering, which is called the Duncan–Mortensen–Zakai equation, or, simply, Zakai's equation. See, e.g., Bensoussan [6] and Xiong [104] for a systematic account.

Most results of Section 3.2 are taken from Wang et al. [95]. If the diffusion coefficients of (2.14) in Section 3.3 contain the state or the control, then (2.14) is not Gaussian in general. Consequently, it is difficult to obtain an explicit filtering equation. The filtering example is taken from Wang and Wu [84], where some applications to optimal control with partially observed information are also studied.

Chapter 3
Optimal Control of Fully Coupled FBSDE with Partial Information

In this chapter, we study an optimal control problem of fully coupled FBSDE with partial information, i.e., Problem A introduced in Section 1.2. Using the convex variation and the duality technique, we derive a stochastic maximum principle and two verification theorems for optimality of Problem A. As an application of the optimality conditions, we solve explicitly an LQ optimal control problem and a cash management problem.

Throughout this chapter, we adopt the following shorthand notations for simplicity:

$$\alpha(t,v) = \alpha(t,x^v(t),y^v(t),z^v(t),v(t)),$$

$$H(t,v) = H(t,x^v(t),y^v(t),z^v(t),v(t);p(t),q(t),k(t)),$$

and

$$\alpha_X(t,u) = \frac{\partial}{\partial X}\alpha(t,u),$$

where we set $\alpha = b,\ \sigma,\ g,\ l,\ H$ and $X = x,\ y,\ z,\ v$.

3.1 Stochastic Maximum Principle

Recall that the state equation for Problem A is given by

$$\begin{cases} dx^v(t) = b(t,v)dt + \sigma(t,v)dW(t), \\ -dy^v(t) = g(t,v)dt - z^v(t)dW(t), \\ x^v(0) = x_0, \quad y^v(T) = f(x^v(T)), \end{cases} \tag{3.1}$$

G. Wang et al., *An Introduction to Optimal Control of FBSDE with Incomplete Information*, SpringerBriefs in Mathematics,
https://doi.org/10.1007/978-3-319-79039-8_3

and the cost functional is

$$J(v(\cdot)) = \mathbb{E}\left[\int_0^T l(t,v)dt + \psi(x^v(T)) + \phi(y^v(0))\right]. \tag{3.2}$$

The information is provided by a sub-σ-field filtration $\mathscr{G}_t \subset \mathscr{F}_t$.

Let $u(\cdot)$ be an optimal control of Problem A, and $(x(\cdot), y(\cdot), z(\cdot))$ the corresponding optimal state. For any given $\varepsilon \in (0,1)$ and v such that $v + u \in \mathscr{U}_{ad}$. By the convexity of $\mathscr{U}_{ad}$, we see that $u + \varepsilon v \in \mathscr{U}_{ad}$. Let $(x^{u+\varepsilon v}(\cdot), y^{u+\varepsilon v}(\cdot), z^{u+\varepsilon v}(\cdot))$ be the solution of (3.1) along with the control $u(\cdot) + \varepsilon v(\cdot)$. Then we have the following continuity dependency of the solution on the control process.

Lemma 3.1. *Under (H1.1)–(H1.3), there is a constant $C > 0$ such that*

$$\mathbb{E}\int_0^T |x^{u+\varepsilon v}(t) - x(t)|^2 dt \le C\varepsilon^2 \mathbb{E}\int_0^T |v(t)|^2 dt,$$
$$\mathbb{E}\int_0^T |y^{u+\varepsilon v}(t) - y(t)|^2 dt \le C\varepsilon^2 \mathbb{E}\int_0^T |v(t)|^2 dt,$$
$$\mathbb{E}\int_0^T |z^{u+\varepsilon v}(t) - z(t)|^2 dt \le C\varepsilon^2 \mathbb{E}\int_0^T |v(t)|^2 dt.$$

Proof. Let

$$\bar{\chi}(\cdot) = \chi^{u+\varepsilon v}(\cdot) - \chi(\cdot),$$

where $\chi = x, y, z$. It is easy to see from (3.1) that

$$\begin{cases} d\bar{x}(t) = [b(t,u+\varepsilon v) - b(t,u)]dt + [\sigma(t,u+\varepsilon v) - \sigma(t,u)]dW(t), \\ -d\bar{y}(t) = [g(t,u+\varepsilon v) - g(t,u)]dt - \bar{z}(t)dW(t), \\ \bar{x}(0) = 0, \quad \bar{y}(T) = f(x^{u+\varepsilon v}(T)) - f(x(T)). \end{cases} \tag{3.3}$$

Applying Itô's formula to $\langle \bar{x}(\cdot), \bar{y}(\cdot)\rangle$, we get

$$\begin{aligned} d\langle \bar{x}(t), \bar{y}(t)\rangle = &\ \langle b(t,u+\varepsilon v) - b(t,u), \bar{y}(t)\rangle dt - \langle g(t,u+\varepsilon v) - g(t,u), \bar{x}(t)\rangle dt \\ &+ \langle \sigma(t,u+\varepsilon v) - \sigma(t,u), \bar{z}(t)\rangle dt + d(mart.). \end{aligned}$$

Taking integration and expectation, we then have

$$\begin{aligned} &\mathbb{E}\langle \bar{x}(T), f(x^{u+\varepsilon v}(T)) - f(x(T))\rangle \\ &= \mathbb{E}\int_0^T \langle b(t,u+\varepsilon v) - b(t,u), \bar{y}(t)\rangle dt \\ &\quad - \mathbb{E}\int_0^T \langle g(t,u+\varepsilon v) - g(t,u), \bar{x}(t)\rangle dt \\ &\quad + \mathbb{E}\int_0^T \langle \sigma(t,u+\varepsilon v) - \sigma(t,u), \bar{z}(t)\rangle dt. \end{aligned} \tag{3.4}$$

Note that

$$
\begin{aligned}
&\langle b(t,u+\varepsilon v)-b(t,u),\bar{y}(t)\rangle \\
&= \langle b(t,u+\varepsilon v)-b(t,x^u(t),y^u(t),z^u(t),u(t)+\varepsilon v(t)),\bar{y}(t)\rangle \\
&\quad + \langle b(t,x^u(t),y^u(t),z^u(t),u(t)+\varepsilon v(t))-b(t,u),\bar{y}(t)\rangle,
\end{aligned}
$$

and by Condition (H1.2),

$$
\langle b(t,x^u(t),y^u(t),z^u(t),u(t)+\varepsilon v(t))-b(t,u),\bar{y}(t)\rangle \le \varepsilon^2 K_\delta |v(t)|^2+\delta|\bar{y}(t)|^2,
$$

where $\delta > 0$ is a constant which could be made arbitrarily small and K_δ is a constant depending on δ.

Similar estimates hold for $-g$ and σ. We can then continue (3.4) with

$$
\begin{aligned}
&\mathbb{E}\left\langle \bar{x}(T), f(x^{u+\varepsilon v}(T))-f(x(T))\right\rangle \\
&\le \mathbb{E}\int_0^T \langle b(t,u+\varepsilon v)-b(t,x^u(t),y^u(t),z^u(t),u(t)+\varepsilon v(t)),\bar{y}(t)\rangle\, dt \\
&\quad +\mathbb{E}\int_0^T \langle -g(t,u+\varepsilon v)+g(t,x^u(t),y^u(t),z^u(t),u(t)+\varepsilon v(t)),\bar{x}(t)\rangle\, dt \\
&\quad +\mathbb{E}\int_0^T \langle \sigma(t,u(t)+\varepsilon v(t))-\sigma(t,x^u(t),y^u(t),z^u(t),u(t)+\varepsilon v(t)),\bar{z}(t)\rangle\, dt \\
&\quad +\mathbb{E}\int_0^T \delta\left(3\varepsilon^2|v(t)|^2+|\bar{x}(t)|^2+|\bar{y}(t)|^2+|\bar{z}(t)|^2\right) dt \\
&\le \mathbb{E}\int_0^T \left(-\mu_1|\bar{x}(t)|^2-\mu_2\left(|\bar{y}(t)|^2+|\bar{z}(t)|^2\right)\right) dt \\
&\quad +\mathbb{E}\int_0^T \left(3\varepsilon^2 K_\delta|v(t)|^2+\delta\left(|\bar{x}(t)|^2+|\bar{y}(t)|^2+|\bar{z}(t)|^2\right)\right) dt,
\end{aligned}
\tag{3.5}
$$

where the last estimate follows from Condition (H1.3). On the other hand, by (H1.3) again, we have

$$
\mathbb{E}\left\langle \bar{x}(T), f(x^{u+\varepsilon v}(T))-f(x(T))\right\rangle \ge \mu_3|\bar{x}(T)|^2. \tag{3.6}
$$

Combining (3.5) and (3.6), we arrive at

$$
\begin{aligned}
&\mu_3\mathbb{E}|\bar{x}(T)|^2+\mu_1\mathbb{E}\int_0^T |\bar{x}(t)|^2 dt+\mu_2\mathbb{E}\int_0^T (|\bar{y}(t)|^2+|\bar{z}(t)|^2)dt \\
&\le \delta\mathbb{E}\int_0^T (|\bar{x}(t)|^2+|\bar{y}(t)|^2+|\bar{z}(t)|^2)dt+3\varepsilon^2 K_\delta\mathbb{E}\int_0^T |v(t)|^2 dt.
\end{aligned}
\tag{3.7}
$$

Further, applying the usual techniques to $|\bar{x}(t)|^2$ and $|\bar{y}(t)|^2$, we get

$$\sup_{0\le t\le T} \mathbb{E}|\bar{x}(t)|^2 \le C_1\mathbb{E}\left[\int_0^T (|\bar{y}(t)|^2+|\bar{z}(t)|^2)dt+\varepsilon^2\int_0^T |v(t)|^2dt\right],$$

$$\mathbb{E}\int_0^T |\bar{x}(t)|^2dt \le C_1\mathbb{E}\left[\int_0^T (|\bar{y}(t)|^2+|\bar{z}(t)|^2)dt+\varepsilon^2\int_0^T |v(t)|^2dt\right],$$

$$\mathbb{E}\int_0^T (|\bar{y}(t)|^2+|\bar{z}(t)|^2)dt \le C_1\mathbb{E}\left[\int_0^T |\bar{x}(t)|^2dt+\varepsilon^2\int_0^T |v(t)|^2dt\right],$$

where C_1 is a constant depending on T.

If $\mu_2>0$, we take $\delta<\frac{\mu_2}{1+K_1}$; if $\mu_2=0$, which implies $\mu_1>0$, we take $\delta<\frac{\mu_1}{1+K_1}$. Combining the above four estimates, it is then clear that we always have

$$\mathbb{E}\int_0^T (|\bar{x}(t)|^2+|\bar{y}(t)|^2+|\bar{z}(t)|^2)dt \le C\varepsilon^2\mathbb{E}\int_0^T |v(t)|^2dt,$$

where C depends on μ_1, μ_2, T, and δ. The proof is then completed. □

Next, we introduce a variational equation

$$\begin{cases} dx^1(t) = \big[b_x(t,u)x^1(t)+b_y(t,u)y^1(t)+b_z(t,u)z^1(t) \\ \qquad\qquad +b_v(t,u)v(t)\big]dt+\big[\sigma_x(t,u)x^1(t)+\sigma_y(t,u)y^1(t) \\ \qquad\qquad +\sigma_z(t,u)z^1(t)+\sigma_v(t,u)v(t)\big]dW(t), \\ -dy^1(t) = \big[g_x(t,u)x^1(t)+g_y(t,u)y^1(t)+g_z(t,u)z^1(t) \\ \qquad\qquad +g_v(t,u)v(t)\big]dt-z^1(t)dW(t), \\ x^1(0)=0, \quad y^1(T)=f_x(x(T))x^1(T), \end{cases} \tag{3.8}$$

which has a unique solution under (H1.1)–(H1.3). As we will see from Lemma 3.2 below, (x^1,y^1,z^1) is the derivation in ε of $(x^{u+\varepsilon v},y^{u+\varepsilon v},z^{u+\varepsilon v})$ at $\varepsilon=0$.

For $X=x,y,z$, let

$$X^\varepsilon(t)=\frac{X^{u+\varepsilon v}(t)-X(t)}{\varepsilon}-X^1(t).$$

Using the arguments similar to Lemma 3.1, we derive the following result.

Lemma 3.2. *If (H1.1)–(H1.3) hold, then*

$$\lim_{\varepsilon\to 0}\mathbb{E}\int_0^T (|x^\varepsilon(t)|^2+|y^\varepsilon(t)|^2+|z^\varepsilon(t)|^2)dt=0.$$

Proof. It follows from (3.3), (3.8), and Taylor's expansion that

$$\begin{cases} dx^\varepsilon(t) = (b_1^\varepsilon(t)x^\varepsilon(t)+b_2^\varepsilon(t)y^\varepsilon(t)+b_3^\varepsilon(t)z^\varepsilon(t)+b_4^\varepsilon(t))\,dt \\ \qquad\qquad +(\sigma_1^\varepsilon(t)x^\varepsilon(t)+\sigma_2^\varepsilon(t)y^\varepsilon(t)+\sigma_3^\varepsilon(t)z^\varepsilon(t)+\sigma_4^\varepsilon(t))\,dW(t), \\ -dy^\varepsilon(t) = (g_1^\varepsilon(t)x^\varepsilon(t)+g_2^\varepsilon(t)y^\varepsilon(t)+g_3^\varepsilon(t)z^\varepsilon(t)+g_4^\varepsilon(t))\,dt - z^\varepsilon(t)dW(t), \\ x^\varepsilon(0)=0, \quad y^\varepsilon(T) = f_1^\varepsilon(T)x^\varepsilon(T)+f_2^\varepsilon(T)x^1(T), \end{cases}$$

where

$$\alpha_X^\varepsilon(t) = \int_0^1 \alpha_X(t,\Upsilon^{\lambda,\varepsilon}(t))d\lambda, \qquad X = x,y,z,$$

$$\begin{aligned} \alpha_4^\varepsilon(t) &= [\alpha_x^\varepsilon(t)-\alpha_x(t,u)]x^1(t)+[\alpha_y^\varepsilon(t)-\alpha_y(t,u)]y^1(t) \\ &\quad +[\alpha_z^\varepsilon(t)-\alpha_z(t,u)]z^1(t)+\left[\int_0^1 \alpha_v(t,\Upsilon^{\lambda,\varepsilon}(t))d\lambda-\alpha_v(t,u)\right]v(t), \end{aligned}$$

$$\begin{aligned} f_1^\varepsilon(T) &= \int_0^1 f_x(x(T)+\lambda(x^{u+\varepsilon v}(T)-x(T)))d\lambda, \\ f_2^\varepsilon(T) &= f_1^\varepsilon(T)-f_x(x(T)) \end{aligned}$$

with $\alpha = b,\sigma,g$, and

$$\begin{aligned} \Upsilon^{\lambda,\varepsilon}(t) = (&x(t)+\lambda(x^{u+\varepsilon v}(t)-x(t)),y(t)+\lambda(y^{u+\varepsilon v}(t)-y(t)), \\ &z(t)+\lambda(z^{u+\varepsilon v}(t)-z(t)),u(t)+\lambda\varepsilon v(t)). \end{aligned}$$

Applying Itô's formula to $|x^\varepsilon(t)|^2$, we get

$$\begin{aligned} \mathbb{E}|x^\varepsilon(t)|^2 \le C_1\mathbb{E}&\int_0^t (|x^\varepsilon(s)|^2+|y^\varepsilon(s)|^2+|z^\varepsilon(s)|^2)ds \\ &+C_1'\mathbb{E}\int_0^t (|b_4^\varepsilon(s)|^2+|\sigma_4^\varepsilon(s)|^2)ds. \end{aligned}$$

Then,

$$\begin{aligned} \sup_{0\le t\le T} \mathbb{E}|x^\varepsilon(t)|^2 \le C_1\mathbb{E}&\int_0^T (|x^\varepsilon(t)|^2+|y^\varepsilon(t)|^2+|z^\varepsilon(t)|^2)dt \\ &+C_1'\mathbb{E}\int_0^T (|b_4^\varepsilon(t)|^2+|\sigma_4^\varepsilon(t)|^2)dt, \end{aligned}$$

and

$$\begin{aligned} \mathbb{E}\int_0^T |x^\varepsilon(t)|^2dt \le C_1T\mathbb{E}&\int_0^T (|x^\varepsilon(t)|^2+|y^\varepsilon(t)|^2+|z^\varepsilon(t)|^2)dt \\ &+C_1'T\mathbb{E}\int_0^T (|b_4^\varepsilon(t)|^2+|\sigma_4^\varepsilon(t)|^2)dt, \end{aligned}$$

where $C_1>0$ and $C_1'>0$ are constants. Similarly, we have

$$\begin{aligned} \mathbb{E}\left(|y^\varepsilon(t)|^2+\int_t^T |z^\varepsilon(s)|^2ds\right) &\le C_2\mathbb{E}\int_t^T (|x^\varepsilon(s)|^2+|y^\varepsilon(s)|^2)ds + C_2'\mathbb{E}|x^\varepsilon(T)|^2 \\ &\quad +C_2''\mathbb{E}\left(|f_2^\varepsilon(T)|^2+\int_t^T |g_4^\varepsilon(s)|^2ds\right). \end{aligned}$$

Hence,

$$\mathbb{E}\int_0^T (|y^\varepsilon(t)|^2 + |z^\varepsilon(t)|^2)dt \leq C_2'(T+2)\mathbb{E}\left[\int_0^T (|x^\varepsilon(t)|^2 + |y^\varepsilon(t)|^2)dt + |x^\varepsilon(T)|^2\right]$$
$$+ C_2''(T+2)\mathbb{E}\left(|f_2^\varepsilon(T)|^2 + \int_0^T |g_4^\varepsilon(t)|^2 dt\right),$$

where $C_2 > 0$, $C_2' > 0$, and $C_2'' > 0$ are constants.

Applying Itô's formula to $\langle x^\varepsilon(\cdot), y^\varepsilon(\cdot)\rangle$ with (H1.3), we derive

$$\begin{aligned}
&\mu_3\mathbb{E}|x^\varepsilon(T)|^2 + \mu_1\mathbb{E}\int_0^T |x^\varepsilon(t)|^2 dt + \mu_2\mathbb{E}\int_0^T (|y^\varepsilon(t)|^2 + |z^\varepsilon(t)|^2)dt \\
&\leq \mathbb{E}\left[\int_0^T (\langle y^\varepsilon(t), b_4^\varepsilon(t)\rangle - \langle x^\varepsilon(t), g_4^\varepsilon(t)\rangle + \langle z^\varepsilon(t), \sigma_4^\varepsilon(t)\rangle)\, dt - \langle x^\varepsilon(T), f_2^\varepsilon(T)\rangle\right] \\
&\leq C_3\mathbb{E}\int_0^T (|x^\varepsilon(t)|^2 + |y^\varepsilon(t)|^2 + |z^\varepsilon(t)|^2)dt \\
&\quad + C_3'\mathbb{E}\left[\int_0^T (|b_4^\varepsilon(t)|^2 + |\sigma_4^\varepsilon(t)|^2 + |g_4^\varepsilon(t)|^2)dt + |f_2^\varepsilon(T)|^2\right],
\end{aligned}$$

where $C_3 > 0$ and $C_3' > 0$ are constants.

Combining the above estimates, we have

$$\begin{aligned}
&\mathbb{E}\int_0^T (|x^\varepsilon(t)|^2 + |y^\varepsilon(t)|^2 + |z^\varepsilon(t)|^2)dt \\
&\leq C_4\mathbb{E}\left[\int_0^T (|b_4^\varepsilon(t)|^2 + |\sigma_4^\varepsilon(t)|^2 + |g_4^\varepsilon(t)|^2)dt + |f_2^\varepsilon(T)|^2\right],
\end{aligned}$$

where $C_4 > 0$ is a constant. Lemma 3.1 and the dominated convergence theorem with (H1.2) imply the desired equality. □

Lemma 3.3. *Under (H1.1)–(H1.5), if $u(\cdot)$ is an optimal control, then for any $v(\cdot)$ such that $v(\cdot) + u(\cdot) \in \mathscr{U}_{ad}$ we have the variational inequality*

$$\begin{aligned}
&\mathbb{E}\left\{\int_0^T \left[l_x^\top(t,u)x^1(t) + l_y^\top(t,u)y^1(t) + l_z^\top(t,u)z^1(t) + l_v^\top(t,u)v(t)\right] dt\right. \\
&\left. + \psi_x^\top(x(T))x^1(T) + \phi_y^\top(y(0))y^1(0)\right\} \geq 0.
\end{aligned} \tag{3.9}$$

Proof. For any $v(\cdot)$ such that $v(\cdot) + u(\cdot) \in \mathscr{U}_{ad}$, it is easy to see

$$\frac{J(u(\cdot) + \varepsilon v(\cdot)) - J(u(\cdot))}{\varepsilon} \geq 0.$$

Applying Taylor's expansion, Lemmas 3.1 and 3.2, we derive easily

$$\lim_{\varepsilon\to 0} \frac{\mathbb{E}[\psi(x^{u+\varepsilon v}(T)) - \psi(x(T))]}{\varepsilon} = \mathbb{E}\left[\psi_x^\top(x(T))x^1(T)\right],$$

$$\lim_{\varepsilon\to 0}\frac{\mathbb{E}[\phi(Y^{u+\varepsilon v}(0))-\phi(y(0))]}{\varepsilon}=\mathbb{E}\left[\phi_y^\top(y(0))y^1(0)\right]$$

and

$$\lim_{\varepsilon\to 0}\frac{\mathbb{E}\int_0^T[l(t,u(t)+\varepsilon v(t))-l(t,u)]dt}{\varepsilon}$$
$$=\mathbb{E}\int_0^T\left[l_x^\top(t,u)x^1(t)+l_y^\top(t,u)y^1(t)+l_z^\top(t,u)z^1(t)+l_v^\top(t,u)v(t)\right]dt.$$

The proof is then completed. □

To obtain a more explicit maximum condition, we recall the adjoint equation

$$\begin{cases} dp(t)= -\left[b_y^\top(t,u)q(t)+\sigma_y^\top(t,u)k(t)-g_y^\top(t,u)p(t)+l_y(t,u)\right]dt \\ \qquad\qquad -\left[b_z^\top(t,u)q(t)+\sigma_z^\top(t,u)k(t)-g_z^\top(t,u)p(t)+l_z(t,u)\right]dW(t), \\ -dq(t)=\left[b_x^\top(t,u)q(t)+\sigma_x^\top(t,u)k(t)-g_x^\top(t,u)p(t)+l_x(t,u)\right]dt-k(t)dW(t), \\ p(0)=-\phi_y(y(0)),\quad q(T)=\psi_x(x(T))-f_x^\top(x(T))p(T). \end{cases} \tag{3.10}$$

If we define a Hamiltonian function $H:[0,T]\times\mathbb{R}^{n+n+n\times m}\times U\times\mathbb{R}^{n+n+n\times m}\to\mathbb{R}$ by

$$\begin{aligned} H(t,x,y,z,v;p,q,k)=&\langle q,b(t,x,y,z,v)\rangle+\langle k,\sigma(t,x,y,z,v)\rangle \\ &-\langle p,g(t,x,y,z,v)\rangle+l(t,x,y,z,v), \end{aligned}$$

then (3.10) is rewritten as

$$\begin{cases} dp(t)=-H_y(t,u)dt-H_z(t,u)dW(t), \\ -dq(t)=H_x(t,u)dt-k(t)dW(t), \\ p(0)=-\phi_y(y(0)),\quad q(T)=\psi_x(x(T))-f_x^\top(x(T))p(T), \end{cases}$$

which is called a generalized stochastic Hamiltonian system.

We now state the maximum principle for Problem A. The backward separation approach appeared here implicitly. Namely, the problems are decoupled by first obtaining the optimal solution and then by solving a filtering equation, which is in reverse order comparing with the usual separation principle. This point will become more clear in Example 3.2 below.

Theorem 3.1. *Under (H1.1)–(H1.5), if $u(\cdot)$ is an optimal control, then (3.10) admits a unique solution $(p(\cdot),q(\cdot),k(\cdot))\in\mathscr{L}_{\mathbb{F}}^2(0,T;\mathbb{R}^{n+n+n\times m})$ such that for any $v\in U$ we have*

$$\langle\mathbb{E}[H_v(t,u)|\mathscr{G}_t],v-u(t)\rangle\geq 0. \tag{3.11}$$

Proof. Applying Itô's formula to $\langle p(\cdot),y^1(\cdot)\rangle+\langle q(\cdot),x^1(\cdot)\rangle$ with (3.8) and (3.10), we derive

$$
\begin{aligned}
&\mathbb{E}\left[\langle x^1(T), \psi_x(x(T))\rangle + \langle y^1(0), \phi_y(y(0))\rangle\right] \\
&= \mathbb{E}\int_0^T [\langle q(t), b_v(t,u)v(t)\rangle + \langle k(t), \sigma_v(t,u)v(t)\rangle + \langle p(t), g_v(t,u)v(t)\rangle]\,dt \\
&\quad - \mathbb{E}\int_0^T \left[\langle x^1(t), l_x(t,u)\rangle + \langle y^1(t), l_y(t,u)\rangle + \langle z^1(t), l_z(t,u)\rangle\right] dt.
\end{aligned}
$$

Inserting it into (3.9), we get

$$
\mathbb{E}\int_0^T \langle \mathbb{E}[H_v(t,u)|\mathscr{G}_t], v(t)\rangle\, dt = \mathbb{E}\int_0^T \langle H_v(t,u), v(t)\rangle dt \geq 0. \tag{3.12}
$$

Now we prove (3.11) by contradiction. Suppose it does not hold, then there exists $\varepsilon > 0$ such that $\mathbb{E}\int_0^T 1_{B_\varepsilon}(t)dt > 0$, where

$$
B_\varepsilon = \{(t,\omega) : \ \langle \mathbb{E}[H_v(t,u)|\mathscr{G}_t], v(t) - u(t)\rangle < -\varepsilon\}.
$$

Define

$$
v^\varepsilon(t) = v(t)1_{B_\varepsilon}(t) + u(t)1_{B_\varepsilon^c}(t) - u(t).
$$

Then,

$$
\begin{aligned}
\mathbb{E}\int_0^T \langle H_v(t,u), v^\varepsilon(t)\rangle dt &= \mathbb{E}\int_0^T \langle H_v(t,u), v(t) - u(t)\rangle 1_{B_\varepsilon}(t)dt \\
&\leq -\varepsilon\mathbb{E}\int_0^T 1_{B_\varepsilon}(t)dt < 0,
\end{aligned}
$$

which contradicts from (3.12), and hence, the proof is completed. □

3.2 Verification Theorem

We now derive some sufficient conditions for the optimality of the controls for Problem A.

Theorem 3.2. *Let (H1.1)–(H1.5) hold. Moreover, we assume that*

- *(Terminal condition) for any* $v(\cdot) \in \mathscr{U}_{ad}$,

$$
y^v(T) = Ax^v(T), \quad A \in \mathbb{R}^{n\times n};
$$

- *(Existence and Uniqueness) for any* $(x(\cdot), y(\cdot), z(\cdot)) \in \mathscr{L}_{\mathbb{F}}^2(0,T;\mathbb{R}^{n+n+n\times m})$ *and* $u(\cdot) \in \mathscr{U}_{ad}$, *FBSDE (3.10) admits a unique solution* $(p(\cdot), q(\cdot), k(\cdot)) \in \mathscr{L}_{\mathbb{F}}^2(0,T;\mathbb{R}^{n+n+n\times m})$;

- *(Minimum Condition) for any* $t \in [0,T]$,

$$\mathbb{E}[H(t,u)|\mathscr{G}_t] = \min_{v \in U} \mathbb{E}[H(t,x(t),y(t),z(t),v;p(t),q(t),k(t))|\mathscr{G}_t];$$

- *(Convexity) for any* $t \in [0,T]$, $H(t,x,y,z,v;p(t),q(t),k(t))$ *is convex in* $(x,y,z,v) \in \mathbb{R}^{n+n+n\times m} \times U$, *and* ψ *and* ϕ *are convex in* $(x,y) \in \mathbb{R}^{n+n}$, *respectively.*

Then $u(\cdot)$ *is an optimal control of Problem A.*

Proof. For any $v(\cdot) \in \mathscr{U}_{ad}$, we consider

$$J(v(\cdot)) - J(u(\cdot)) = I_1 + I_2 + I_3 \tag{3.13}$$

with

$$\begin{aligned}
I_1 &= \mathbb{E}[\phi(y^v(0)) - \phi(y(0))], \\
I_2 &= \mathbb{E}[\psi(x^v(T)) - \psi(x(T))], \\
I_3 &= \mathbb{E}\int_0^T [l(t,v) - l(t,u)]\,dt.
\end{aligned}$$

Recall that ϕ is convex in y and $p(0) = -\phi_y(y(0))$. Applying Itô's formula to $\langle p(t), y^v(t) - y(t)\rangle$, we get

$$\begin{aligned}
I_1 &\geq -\mathbb{E}\left[p^\top(0)(y^v(0) - y(0))\right] \\
&= -\mathbb{E}\left[p^\top(T)A(x^v(T) - x(T))\right] \\
&\quad -\mathbb{E}\int_0^T \langle H_y(t,u), y^v(t) - y(t)\rangle dt - \mathbb{E}\int_0^T \langle H_z(t,u), z^v(t) - z(t)\rangle dt \\
&\quad -\mathbb{E}\int_0^T \langle p(t), g(t,v) - g(t,u)\rangle dt.
\end{aligned} \tag{3.14}$$

Similarly,

$$I_2 \geq \mathbb{E}\left[(q^\top(T) + p^\top(T)A)(x^v(T) - x(T))\right] \tag{3.15}$$

with

$$\begin{aligned}
&\mathbb{E}\left[q^\top(T)(x^v(T) - x(T))\right] \\
&= \mathbb{E}\int_0^T \langle q(t), b(t,v) - b(t,u)\rangle dt + \mathbb{E}\int_0^T \langle k(t), \sigma(t,v) - \sigma(t,u)\rangle dt \\
&\quad -\mathbb{E}\int_0^T \langle H_x(t,u), x^v(t) - x(t)\rangle dt.
\end{aligned}$$

According to (3.13)–(3.15) and I_3, it is easy to see from the convexity of the function $H(t,x,y,z,v;p(t),q(t),k(t))$ that

$$\begin{aligned}J(v(\cdot))-J(u(\cdot)) \geq \mathbb{E}\int_0^T (H(t,v)-H(t,u))\,dt - \mathbb{E}\int_0^T \langle H_x(t,u), x^v(t)-x(t)\rangle dt\\ -\mathbb{E}\int_0^T \langle H_y(t,u), y^v(t)-y(t)\rangle dt - \mathbb{E}\int_0^T \langle H_z(t,u), z^v(t)-z(t)\rangle dt\\ \geq \mathbb{E}\int_0^T \mathbb{E}[\langle H_v(t,u), v(t)-u(t)\rangle|\mathscr{G}_t]dt.\end{aligned}$$

Further, using the minimum condition, we get

$$\left(\frac{\partial}{\partial v}\mathbb{E}[H(t,u)|\mathscr{G}_t]\right)^\top (v(t)-u(t)) = \mathbb{E}[\langle H_v(t,u), v(t)-u(t)\rangle|\mathscr{G}_t] \geq 0.$$

This implies the desired result. □

If $\mathscr{G}_t = \mathscr{F}_t$, $t\in[0,T]$, Theorem 3.2 is reduced to a Mangasarian sufficient condition for optimality. Further, the convexity condition of $H(t,x,y,z,v;p(t),q(t),k(t))$ with respect to (x,y,z,v) can be relaxed, and a generalized verification theorem can be derived. This point is supported by Theorem 3.3 below, which is also called an Arrow sufficient condition for optimality.

Theorem 3.3. *Let (H1.1)–(H1.5) hold. Moreover, we assume that*

- *(Terminal condition) for any* $v(\cdot)\in\mathscr{U}_{ad}$,

$$y^v(T) = Ax^v(T), \quad A\in\mathbb{R}^{n\times n};$$

- *(Existence and Uniqueness)* $(x(\cdot),y(\cdot),z(\cdot))\in\mathscr{L}^2_{\mathbb{F}}(0,T;\mathbb{R}^{n+n+n\times m})$ *and* $u(\cdot)\in\mathscr{U}_{ad}$, *(3.10) admits a unique solution* $(p(\cdot),q(\cdot),k(\cdot))\in\mathscr{L}^2_{\mathbb{F}}(0,T;\mathbb{R}^{n+n+n\times m})$;
- *(Minimum Condition) for any* $t\in[0,T]$,

$$H(t,u) = \min_{v\in U} H(t,x(t),y(t),z(t),v;p(t),q(t),k(t));$$

- *(Convexity) the function*

$$\tilde{H}(t,x,y,z) = \min_{v\in U} H(t,x,y,z,v;p(t),q(t),k(t))$$

exists and is convex in (x,y,z), *and* ψ *and* ϕ *are convex in* x *and* y, *respectively.*

Then $u(\cdot)$ *is an optimal control of Problem A.*

Proof. Similar to the proof of Theorem 3.2, for any $v(\cdot)\in\mathscr{U}_{ad}$ we have

$$
\begin{aligned}
J(v(\cdot)) - J(u(\cdot)) \geq \mathbb{E}\int_0^T (H(t,v) - H(t,u))\,dt \\
-\mathbb{E}\int_0^T \langle H_x(t,u), x^v(t) - x(t)\rangle dt \\
-\mathbb{E}\int_0^T \langle H_y(t,u), y^v(t) - y(t)\rangle dt \\
-\mathbb{E}\int_0^T \langle H_z(t,u), z^v(t) - z(t)\rangle dt.
\end{aligned} \tag{3.16}
$$

Now we show that $J(v(\cdot)) - J(u(\cdot)) \geq 0$ holds for any $v(\cdot) \in \mathscr{U}_{ad}$.

According to the minimum condition, for any (t,x,y,z)

$$
\begin{aligned}
&\tilde{H}(t,x,y,z) - \tilde{H}(t,x(t),y(t),z(t)) \\
&\leq H(t,x,y,z,v;p(t),q(t),k(t)) - H(t,u).
\end{aligned} \tag{3.17}
$$

Fix $t \in [0,T]$. Since $\tilde{H}(t,x,y,z)$ is convex in (x,y,z), there exist $a(t)$, $b(t) \in \mathbb{R}^n$ and $c(t) \in \mathbb{R}^{n\times m}$ such that

$$
\begin{aligned}
\tilde{H}(t,x,y,z) - \tilde{H}(t,x(t),y(t),z(t)) \geq \langle a(t), x - x(t)\rangle + \langle b(t), y - y(t)\rangle \\
+\langle c(t), z - z(t)\rangle.
\end{aligned} \tag{3.18}
$$

Define

$$
\begin{aligned}
\varphi(t,x,y,z) = H(t,x,y,z,u(t);p(t),q(t),k(t)) - H(t,u) \\
-\langle a(t), x - x(t)\rangle + \langle b(t), y - y(t)\rangle + \langle c(t), z - z(t)\rangle.
\end{aligned}
$$

It follows from (3.17) and (3.18) that $\varphi(t,x,y,z) \geq 0$ for all (t,x,y,z). Moreover, $\varphi(t,x(t),y(t),z(t)) = 0$. It implies that $(x(\cdot),y(\cdot),z(\cdot))$ is a minimum point of φ. Because φ is differentiable with respect to (x,y,z), we have the partial derivatives

$$
\varphi_x(t,x(t),y(t),z(t)) = 0, \quad \varphi_y(t,x(t),y(t),z(t)) = 0, \quad \varphi_z(t,x(t),y(t),z(t)) = 0,
$$

i.e.,

$$
H_x(t,u) = a(t), \quad H_y(t,u) = b(t), \quad H_z(t,u) = c(t).
$$

Inserting them into (3.18), we see that

$$
\begin{aligned}
\tilde{H}(t,x,y,z) - \tilde{H}(t,x(t),y(t),z(t)) \geq \langle H_x(t,u), x - x(t)\rangle + \langle H_y(t,u), y - y(t)\rangle \\
+\langle H_z(t,u), z - z(t)\rangle.
\end{aligned} \tag{3.19}
$$

Combing with (3.16) and (3.17) we see that $J(v(\cdot)) \geq J(u(\cdot))$, i.e., $u(\cdot)$ is an optimal control. The proof is then completed. □

In the rest of this section, we work out an example to show that the Arrow sufficient condition is really needed.

Example 3.1. Consider a control system ($n = m = k = 1$)

$$\begin{cases} dx(t) = v(t)dt + dW(t), \\ -dy(t) = v(t)dt - z(t)dW(t), \\ x(0) = x_0, \quad y(T) = x(T), \end{cases}$$

with $U = [0, \infty)$ and

$$J(v(\cdot)) = \mathbb{E}\left\{\int_0^T \min\left\{v^2(t) - v(t), 1\right\} dt + 3x(T) - y(0)\right\}.$$

Since the diffusion coefficient is a constant and the drift coefficients do not depend on the state explicitly, the Hamiltonian function and the adjoint equation are reduced to

$$H(t,x,y,z,v;p,q,k) = \min\left\{v^2 - v, 1\right\} + qv + k - pv$$

and

$$\begin{cases} \dot{p}(t) = 0, \quad p(0) = 1, \\ dq(t) = k(t)dW(t), \quad q(T) = 3 - p(T). \end{cases}$$

Solving the FBSDE above, we get $p(t) \equiv 1$, $q(t) \equiv 2$, and $k(t) \equiv 0$. Substituting it into the Hamiltonian function, we have

$$\begin{aligned} H(t,x,y,z,v;p(t),q(t),k(t)) &= \min\left\{v^2 - v, 1\right\} + v \\ &= \begin{cases} v^2, & \text{if } v \in [0, (1+\sqrt{5})/2), \\ v+1, & \text{if } v \in [(1+\sqrt{5})/2, \infty), \end{cases} \end{aligned}$$

which is neither a convex nor a concave function of v on the whole horizon $[0, \infty)$. It is easy to see that $H(t,x,y,z,v;p(t),q(t),k(t))$ attains its minimum value 0 at $v = 0$, i.e., $\tilde{H}(t,x,y,z) \equiv 0$. On the other hand, suppose that $u(t) \equiv 0$. Clearly,

$$H(t,u) \equiv \min\{0^2 - 0, 1\} + 0 = 0.$$

Then

$$H(t,u) = \min_{v \in U} H(t,x(t),y(t),z(t),v;p(t),q(t),k(t)) \equiv 0.$$

Now all the assumptions required in Theorem 3.3 are satisfied, and hence, $u(t) \equiv 0$ is an optimal control of Example 3.1.

3.3 An LQ Optimal Control Problem

The aim of this section is to demonstrate the applications of the theoretical results obtained above via an LQ problem. Although the cost functional looks simple, the mathematical deductions used to find an optimal control are still nontrivial. One main motivation for this example is that the observable data is not complete, and it is necessary to compute optimal estimates of FBSDEs based on the observable information. However, these estimates are infinite dimensional, in general.

Example 3.2. Consider a partial information LQ problem with $U = \mathbb{R}$:

$$\min_{v(\cdot)\in\mathscr{U}_{ad}} J(v(\cdot))$$

subject to

$$J(v(\cdot)) = \mathbb{E}\left[\frac{1}{2}\int_0^T v^2(t)dt + y^v(0)\right]$$

and a fully coupled FBSDE

$$\begin{cases} dx(t) = (a_1x(t)+a_2y(t)+a_3z_1(t)+a_4z_2(t)+a_5v(t))dt \\ \qquad\qquad + (b_1x(t)+b_2y(t)+b_3z_1(t)+b_4z_2(t)+b_5v(t))dW_1(t) \\ \qquad\qquad + (c_1x(t)+c_2y(t)+c_3z_1(t)+c_4z_2(t)+c_5v(t))dW_2(t), \\ -dy(t) = (e_1x(t)+a_1y(t)+b_1z_1(t)+c_1z_2(t)+e_2v(t))dt \\ \qquad\qquad - z_1(t)dW_1(t) - z_2(t)dW_2(t), \\ x(0) = x_0, \quad y(T) = x(T). \end{cases} \tag{3.20}$$

For the monotonicity to be satisfied, we make the following assumptions on the constants: $a_2 < 0$, $b_2 = -a_3$, $b_3 < 0$, $c_2 = -a_4$, $c_3 = -b_4$, $c_4 < 0$, and $e_1 > 0$. Namely, for any $v(\cdot) \in \mathscr{U}_{ad}$, we can check that (H1.1)–(H1.3) are satisfied; then, (3.20) has a unique solution $(x^v(\cdot), y^v(\cdot), z_1^v(\cdot), z_2^v(\cdot)) \in \mathscr{L}_{\mathbb{F}}^2(0,T;\mathbb{R}^4)$.

In this situation, the adjoint equation is

$$\begin{cases} dp(t) = (a_1p(t)-a_2q(t)+a_3k_1(t)+a_4k_2(t))dt \\ \qquad\qquad + (b_1p(t)-a_3q(t)-b_3k_1(t)+b_4k_2(t))dW_1(t) \\ \qquad\qquad + (c_1p(t)-a_4q(t)-b_4k_1(t)-c_4k_2(t))dW_2(t), \\ -dq(t) = (a_1q(t)+b_1k_1(t)+c_1k_2(t)-e_1p(t))dt \\ \qquad\qquad - k_1(t)dW_1(t)-k_2(t)dW_2(t), \\ p(0) = -1, \quad q(T) = -p(T). \end{cases} \tag{3.21}$$

Since (H1.1), (H1.2), and (H1.3)' are applicable to (3.21), it admits a unique solution $(p(\cdot), q(\cdot), k_1(\cdot), k_2(\cdot)) \in \mathscr{L}_{\mathbb{F}}^2(0,T;\mathbb{R}^4)$. Let

$$u(t) = e_2\mathbb{E}[p(t)|\mathscr{G}_t] - a_5\mathbb{E}[q(t)|\mathscr{G}_t] - b_5\mathbb{E}[k_1(t)|\mathscr{G}_t] - c_5\mathbb{E}[k_2(t)|\mathscr{G}_t]. \tag{3.22}$$

Clearly, $u(\cdot)$ is in $\mathscr{U}_{ad}$. The Hamiltonian function is

$$\begin{aligned} H(t,x,y,z_1,z_2,v;p,q,k_1,k_2) = {} & \frac{1}{2}v^2 + (a_1x+a_2y+a_3z_1+a_4z_2+a_5v)q \\ & + (b_1x-a_3y+b_3z_1+b_4z_2+b_5v)k_1 \\ & + (c_1x-a_4y-b_4z_1+c_4z_2+c_5v)k_2 \\ & - (e_1x+a_1y+b_1z_1+c_1z_2+e_2v)p. \end{aligned}$$

Then

$$
\begin{aligned}
&\mathbb{E}[H(t,x(t),y(t),z_1(t),z_2(t),v;p(t),q(t),k_1(t),k_2(t))|\mathscr{G}_t]\\
&=\frac{1}{2}[v-(e_2\mathbb{E}[p(t)|\mathscr{G}_t]-a_5\mathbb{E}[q(t)|\mathscr{G}_t]-b_5\mathbb{E}[k_1(t)|\mathscr{G}_t]-c_5\mathbb{E}[k_2(t)|\mathscr{G}_t])]^2\\
&\quad-\frac{1}{2}(e_2\mathbb{E}[p(t)|\mathscr{G}_t]-a_5\mathbb{E}[q(t)|\mathscr{G}_t]-b_5\mathbb{E}[k_1(t)|\mathscr{G}_t]-c_5\mathbb{E}[k_2(t)|\mathscr{G}_t])^2\\
&\quad+\mathbb{E}[(a_1q(t)+b_1k_1(t)+c_1k_2(t)-e_1p(t))x(t)|\mathscr{G}_t]\\
&\quad+\mathbb{E}[(a_2q(t)-a_3k_1(t)-a_4k_2(t)-a_1p(t))y(t)|\mathscr{G}_t]\\
&\quad+\mathbb{E}[(a_3q(t)+b_3k_1(t)-b_4k_2(t)-b_1p(t))z_1(t)|\mathscr{G}_t]\\
&\quad+\mathbb{E}[(a_4q(t)+b_4k_1(t)+c_4k_2(t)-c_1p(t))z_2(t)|\mathscr{G}_t].
\end{aligned}
$$

Furthermore, the function $(x,y,z_1,z_2,v)\to H(t,x,y,z_1,z_2,v;p(t),q(t),k_1(t),k_2(t))$ is convex. Therefore it follows from Theorem 3.2 that $u(\cdot)$, defined by (3.22), is an optimal control.

The explicit form of (3.22) is rarely available except for some special settings. In what follows, we will obtain explicit optimal controls by optimal filtering of FBSDEs for these special cases. We also need an additional assumption.

(H3.1) $a_3=a_4=b_3=b_4=c_4=0$ and $\mathscr{G}_t=\sigma\{W_1(s):\ 0\le s\le t\}$.

Based on (H3.1), (3.21) is reduced to

$$
\begin{cases}
dp(t)=(a_1p(t)-a_2q(t))dt+b_1p(t)dW_1(t)+c_1p(t)dW_2(t),\\
-dq(t)=(a_1q(t)+b_1k_1(t)+c_1k_2(t)-e_1p(t))dt\\
\qquad\qquad -k_1(t)dW_1(t)-k_2(t)dW_2(t),\\
p(0)=-1,\quad q(T)=-p(T).
\end{cases}
\tag{3.23}
$$

Applying Theorems 3.1 and 3.2 to (3.23), we obtain the filtering equation

$$
\begin{cases}
d\hat{p}(t)=(a_1\hat{p}(t)-a_2\hat{q}(t))dt+b_1\hat{p}(t)dW_1(t),\\
-d\hat{q}(t)=(a_1\hat{q}(t)+b_1\hat{k}_1(t)+c_1\hat{k}_2(t)-e_1\hat{p}(t))dt-\hat{k}_1(t)dW_1(t),\\
\hat{p}(0)=-1,\quad \hat{q}(T)=-\hat{p}(T).
\end{cases}
\tag{3.24}
$$

This is referred to as a kind of fully coupled forward-backward stochastic differential filtering equation. Note that, since $\hat{k}_2(\cdot)$ appears in the BSDE, (3.24) is not a standard FBSDE, and consequently, the existence and uniqueness of its solution is not an immediate conclusion.

Noting the terminal condition of (3.23), we set

$$
q(\cdot)=\pi(\cdot)p(\cdot)
\tag{3.25}
$$

with $\pi(T)=-1$, where $\pi(\cdot)$ is deterministic and will be given later on. Applying Itô's formula to (3.25),

$$
\begin{aligned}
dq(t)=&\ [\dot{\pi}(t)p(t)+\pi(t)(a_1p(t)-a_2q(t))]dt\\
&+b_1\pi(t)p(t)dW_1(t)+c_1\pi(t)p(t)dW_2(t).
\end{aligned}
$$

Comparing it with the BSDE in (3.23), we derive

$$k_1(t) = b_1\pi(t)p(t), \quad k_2(t) = c_1\pi(t)p(t), \tag{3.26}$$

and

$$\begin{aligned}&\dot{\pi}(t)p(t) + \pi(t)(a_1 p(t) - a_2 q(t))\\ &= e_1 p(t) - a_1 q(t) - b_1 k_1(t) - c_1 k_2(t).\end{aligned} \tag{3.27}$$

Inserting (3.25) and (3.26) into (3.27), we obtain

$$\begin{cases} \dot{\pi}(t) + (2a_1 + b_1^2 + c_1^2)\pi(t) - a_2\pi^2(t) - e_1 = 0, \\ \pi(T) = -1. \end{cases} \tag{3.28}$$

Recall that $a_2 < 0$ and $e_1 > 0$. Then (3.28) admits a unique solution $\pi(t) \le 0$. Taking conditional expectations with respect to $\mathscr{G}_t$ on both sides of (3.26),

$$\hat{k}_1(t) = b_1\pi(t)\hat{p}(t), \quad \hat{k}_2(t) = c_1\pi(t)\hat{p}(t). \tag{3.29}$$

Substituting the second equality of (3.29) into (3.24), we arrive at

$$\begin{cases} d\hat{p}(t) = (a_1\hat{p}(t) - a_2\hat{q}(t))dt + b_1\hat{p}(t)dW_1(t), \\ -d\hat{q}(t) = \left[a_1\hat{q}(t) + b_1\hat{k}_1(t) + (c_1^2\pi(t) - e_1)\hat{p}(t)\right]dt - \hat{k}_1(t)dW_1(t), \\ \hat{p}(0) = -1, \quad \hat{q}(T) = -\hat{p}(T). \end{cases} \tag{3.30}$$

According to (H1.1), (H1.2), and (H1.3)', the filtering equation (3.30) admits a unique solution $(\hat{p}(\cdot), \hat{q}(\cdot), \hat{k}_1(\cdot)) \in \mathscr{L}^2_{\mathbb{G}}(0,T;\mathbb{R}^3)$. Therefore, the optimal control is

$$\begin{aligned} u(t) &= [e_2 - (a_5 + b_1b_5 + c_1c_5)\pi(t)]\hat{p}(t) \\ &= [(a_5 + b_1b_5 + c_1c_5)\pi(t) - e_2]e^{\int_0^t (a_1 - a_2\pi(s) - \frac{1}{2}b_1^2)ds + b_1W_1(t)}, \end{aligned} \tag{3.31}$$

where the second equality follows by solving (3.30).

We conclude the above analysis as follows.

Proposition 3.1. *The optimal control of Example 3.2 is given by (3.22). In addition, if (H3.1) holds, then the optimal control is explicitly expressed by (3.31).*

This result extends partially some literature, for example, [54, 61].

3.4 A Cash Management Problem

Cash management has important values in both theoretical and practical aspects. However, such a class of problems has often been ignored in literature. Let us consider a fully coupled FBSDE ($n = m = k = 1$)

$$\begin{cases} dx(t) = (-\alpha_1 x(t) - \beta_1 y(t) + \gamma_1 v(t))dt - \beta_2 z(t)dW(t), \\ -dy(t) = (-\alpha_1 y(t) + \alpha_2 x(t) + \gamma_2 v(t))dt - z(t)dW(t), \\ x(0) = x_0, \quad y(T) = ax(T), \end{cases} \tag{3.32}$$

where constants $x_0 \in \mathbb{R}$, $a > 0$, $\alpha_i > 0$, $\beta_i > 0$, $\gamma_i \in \mathbb{R} - \{0\}$ $(i = 1,2)$. $x(\cdot)$ describes a cash flow of an agent, $v(\cdot)$ is a control strategy of the agent and is interpreted as the rate of capital injection or withdrawal, so as to achieve a goal. $y(\cdot)$ denotes the utility from $v(\cdot)$, while $z^2(\cdot)$ is referred as the utility "volatility." The terms $-\beta_1 y(t)dt$ and $-\beta_2 z(t)dW(t)$ express the influence of the utility and its volatility on the cash flow. From equation (3.32), we can see intuitively that $y(\cdot)$ is increasing with respect to the cash $x(\cdot)$, while $x(\cdot)$ is decreasing in $y(\cdot)$.

Suppose that the available information is complete. For any $v(\cdot) \in \mathscr{U}_{ad}$, (3.32) has a unique solution $(x^v(\cdot), y^v(\cdot), z^v(\cdot)) \in \mathscr{L}^2_{\mathbb{F}}(0,T;\mathbb{R}^3)$. Introduce the performance functional

$$J(v(\cdot)) = \mathbb{E}\left[\frac{1}{2}\int_0^T (v(t) - b(t))^2 dt - y^v(0)\right],$$

where $b(\cdot)$ is a deterministic and bounded function taking values in $\mathbb{R}$, and is interpreted as a dynamic benchmark. Then the cash management problem with stochastic recursive utility is as follows.

Problem (CM). Find a control strategy $u(\cdot) \in \mathscr{U}_{ad}$ such that

$$J(u(\cdot)) = \min_{v(\cdot)\in\mathscr{U}_{ad}} J(v(\cdot))$$

subject to (3.32). This problem applies to an agent who wants not only to prevent the control strategy from large deviation, but also to maximize the recursive utility.

Recently, Bensoussan et al. [8] solved a cash management problem with random gains rate of stock. [75] studied a mean-variance portfolio selection problem with recursive utility. However, neither [8] nor [75] considered the influence of the recursive utility and its volatility on the cash flow.

Obviously, Problem (CM) is a special case of Problem A. In this case, the Hamiltonian function and the adjoint equation are reduced to

$$\begin{aligned} H(t,x,y,z,v;p,q,k) = {} & \frac{1}{2}(v - b_t - \gamma_2 p + \gamma_1 q)^2 - \gamma_1^2 q^2 - \gamma_2^2 p^2 - \beta_2 zk \\ & -(\alpha_1 x + \beta_1 y - 2\gamma_1 b_t)q - (\alpha_2 x - \alpha_1 y + 2\gamma_2 b_t)p \end{aligned}$$

and

$$\begin{cases} dp(t) = (\beta_1 q(t) - \alpha_1 p(t))dt + \beta_2 k(t)dW(t), \\ -dq(t) = (-\alpha_2 p(t) - \alpha_1 q(t))dt - k(t)dW(t), \\ p(0) = 1, \quad q(T) = -ap(T). \end{cases} \tag{3.33}$$

According to (H1.1), (H1.2), and (H1.3)', FBSDE (3.33) admits a unique solution $(p(\cdot), q(\cdot), k(\cdot)) \in \mathscr{L}^2_{\mathbb{F}}(0,T;\mathbb{R}^3)$.

Let

$$u(t) = b(t) - \gamma_1 q(t) + \gamma_2 p(t). \tag{3.34}$$

It is easy to check that

$$\begin{aligned} H(t,u) &= \min_{v\in U} H(t,x(t),y(t),z(t),v;p(t),q(t),k(t)) \\ &= -\gamma_1^2 q^2(t) - \gamma_2^2 p^2(t) - \beta_2 z(t)k(t) \\ &\quad -(\alpha_1 x(t) + \beta_1 y(t) - 2\gamma_1 b_t)q(t) \\ &\quad -(\alpha_2 x(t) - \alpha_1 y(t) + 2\gamma_2 b_t)p(t). \end{aligned}$$

Moreover,

$$\begin{aligned} \tilde{H}(t,x,y,z) &= H(t,x,y,z,u(t);p(t),q(t),k(t)) \\ &= \min_{v\in U} H(t,x,y,z,v;p(t),q(t),k(t)) \end{aligned}$$

exists and is convex with respect to (x,y,z). Then, by Theorem 3.3 we have

Proposition 3.2. *The optimal strategy of Problem (CM) is given by (3.34), where $(p(\cdot),q(\cdot),k(\cdot))$ solves (3.33).*

3.5 Notes

Maximum principle is a necessary condition of optimal control for an optimization problem. The necessary condition becomes sufficient under some convexity assumptions on cost functional. In 1950s, Pontryagin and his school [70] formulated and derived a milestone result in optimal control theory, i.e., the maximum principle for deterministic systems. The result shows that it is necessary for an optimal control along with the optimal state to satisfy an extended Hamiltonian system, which is a forward-backward differential equation. The earliest papers concerning sufficiency of the maximum principle were published by Mangasarian [53] and Arrow and Kurt [1] with some convex conditions. Shortly after the Pontryagin's maximum principle was obtained, Kushner and Schweppe [42] derived the first stochastic analog for the stochastic diffusion case. The celebrated paper of Peng [66] brings the research of the stochastic maximum principle to a culmination, where the control system is possibly degenerate with a non-convex control domain and a control-dependent diffusion. This is the so-called general stochastic maximum principle. The sufficiency of the general stochastic maximum principle was proved by Zhou [116]. As for the fully coupled forward-backward stochastic control system, how to derive a general maximum principle remains open. The first paper on this topic was published in 1993 by Peng [67], where the SDE and the BSDE are decoupled with a convex control domain and a control-dependent diffusion. After that, lots of attempts were made to extend Peng [67] to a more general case. We mention only a few papers: Xu [106], Wu [99], Shi and Wu [75], Yong [107], Wu [101], Hu [26].

Another often used method to solving optimal control problems is dynamic programming, which was initiated by Bellman [3] when he studied multistage decision problems. It seems that the first stochastic version of dynamic programming in continuous time was obtained by Kushner [41]. Since then, numerous works related to this subject have sprung up. See, e.g., Yong and Zhou [109], Peng [63, 65], Wu and Yu [102], Touzi [81], and literature cited therein. The relationship between the maximum principle and the dynamic programming principle is summarized in Yong and Zhou [109]. See also Nie et al. [58] for recent development. The third alternative method to optimal control is duality theory, which was formulated initially to study a certain utility maximization problem in a complete market. It was then extended to the case of incomplete market. We refer the reader to Pham [69] for a systematic account about the method. See also Bouchard et al. [13] for further development.

The results presented in this chapter are taken mainly from Wu [99] and Wang and Xiao [90]. When the dimension of $x(\cdot)$ is different from that of $y(\cdot)$, Problem A can also be studied in a way similar to Sections 3.1 and 3.2. We omit the details here to save space.

Chapter 4
Optimal Control of FBSDE with Partially Observable Information

In this chapter, we study an optimal control problem with state process governed by a nonlinear FBSDE and with partially observable information, i.e., Problem B introduced in Section 1.2. For simplicity, we take the dimensions $n = m = k = \tilde{k} = 1$. Using a direct method and a Malliavin derivative method, we establish two versions of the stochastic maximum principle for the characterization of the optimal control. To demonstrate the applicability, we work out an illustrative example within the framework of recursive utility and then solve it via the stochastic maximum principle and the stochastic filtering.

4.1 A Direct Method

4.1.1 Some Prior Estimates

Recall that Problem B consists of the state equation

$$\begin{cases} dx^v(t) = & b(t,x^v(t),v(t))dt + \sigma(t,x^v(t),v(t))dW(t) \\ & \quad +\tilde{\sigma}(t,x^v(t),v(t))d\tilde{W}^v(t), \\ -dy^v(t) = & g(t,x^v(t),y^v(t),z^v(t),\tilde{z}^v(t),v(t))dt \\ & \quad -z^v(t)dW(t) - \tilde{z}^v(t)dY(t), \\ x^v(0) = & x_0, \quad y^v(T) = f(x^v(T)), \end{cases} \tag{4.1}$$

and the cost function

$$J(v(\cdot)) = \mathbb{E}^v\left[\int_0^T l(t,x^v(t),y^v(t),z^v(t),\tilde{z}^v(t),v(t))dt + \phi(x^v(T)) + \gamma(y^v(0))\right]. \tag{4.2}$$

G. Wang et al., *An Introduction to Optimal Control of FBSDE with Incomplete Information*, SpringerBriefs in Mathematics,
https://doi.org/10.1007/978-3-319-79039-8_4

The information is provided by the observation equation

$$\begin{cases} dY(t) = h(t,x^v(t))dt + d\tilde{W}^v(t), \\ Y(0) = 0, \end{cases} \tag{4.3}$$

Recall also that the process $Z^v(t)$ is given by (1.16) which helps to transfer $\tilde{W}^v$ into a Brownian motion under a new probability measure $\mathbb{P}^v$.

Let $\varepsilon \in (0,1)$ and $v(\cdot)$ such that $v(\cdot)+u(\cdot) \in \mathscr{U}_{ad}$. By the convexity of U, $u+\varepsilon v \in \mathscr{U}_{ad}$. Denoted by $(x^{u+\varepsilon v}(\cdot), y^{u+\varepsilon v}(\cdot), z^{u+\varepsilon v}(\cdot), \tilde{z}^{u+\varepsilon v}(\cdot))$ and $Z^{u+\varepsilon v}(\cdot)$ the solutions of (4.1) and (1.16) along with the control $u(\cdot)+\varepsilon v(\cdot)$. Making use of the Burkholder–Davis–Gundy (BDG) inequality and Gronwall's inequality, we get the following estimates.

Lemma 4.1. *Under (H1.6), for any $v(\cdot) \in \mathscr{U}_{ad}$ there is a constant $C > 0$ such that the solutions of (1.15) and (1.16) satisfy*

$$\sup_{0\le t\le T} \mathbb{E}|x^v(t)|^8 \le C\left(1 + \sup_{0\le t\le T} \mathbb{E}|v(t)|^8\right),$$

$$\sup_{0\le t\le T} \mathbb{E}|y^v(t)|^2 \le C\left(1 + \sup_{0\le t\le T} \mathbb{E}|v(t)|^2\right),$$

$$\mathbb{E}\left(\int_0^T |z^v(t)|^2 dt + \int_0^T |\tilde{z}^v(t)|^2 dt\right) \le C\left(1 + \sup_{0\le t\le T} \mathbb{E}|v(t)|^2\right),$$

$$\mathbb{E}|Z^v(t)|^\ell < \infty, \quad \forall \ell > 0.$$

Lemma 4.2. *Under (H1.6), there is a constant $C > 0$ such that*

$$\sup_{0\le t\le T} \mathbb{E}|x^{u+\varepsilon v}(t) - x(t)|^8 \le C\varepsilon^8, \quad \sup_{0\le t\le T} \mathbb{E}|y^{u+\varepsilon v}(t) - y(t)|^2 \le C\varepsilon^2,$$

$$\mathbb{E}\int_0^T |z^{u+\varepsilon v}(t) - z(t)|^2 dt \le C\varepsilon^2, \quad \mathbb{E}\int_0^T |\tilde{z}^{u+\varepsilon v}(t) - \tilde{z}(t)|^2 dt \le C\varepsilon^2,$$

$$\sup_{0\le t\le T} \mathbb{E}|Z^{u+\varepsilon v}(t) - Z(t)|^2 \le C\varepsilon^2.$$

We introduce the variational equations

$$\begin{cases} dZ^1(t) = \left(Z^1(t)h(t,x(t)) + Z(t)h_x(t,x(t))x^1(t)\right) dY(t), \\ Z^1(0) = 0 \end{cases} \tag{4.4}$$

and

$$
\begin{cases}
dx^1(t) = \{[b_x(t,u) - \tilde{\sigma}_x(t,u)h(t,x(t)) - \tilde{\sigma}(t,u)h_x(t,x(t))]x^1(t) \\
\qquad + [b_v(t,u) - \tilde{\sigma}_v(t,u)h(t,x(t))]v(t)\}dt \\
\qquad + \left[\sigma_x(t,u)x^1(t) + \sigma_v(t,u)v(t)\right]dW(t) \\
\qquad + \left[\tilde{\sigma}_x(t,u)x^1(t) + \tilde{\sigma}_v(t,u)v(t)\right]dY(t), \\
-dy^1(t) = \left[g_x(t,u)x^1(t) + g_y(t,u)y^1(t) + g_z(t,u)z^1(t)\right. \\
\qquad \left. + g_{\tilde{z}}(t,u)\tilde{z}^1(t) + g_v(t,u)v(t)\right]dt \\
\qquad - z^1(t)dW(t) - \tilde{z}^1(t)dY(t), \\
x^1(0) = 0, \quad y^1(T) = f_x(x(T))x^1(T),
\end{cases}
\tag{4.5}
$$

where we used the notation convention of the last chapter. For example,

$$b_x(t,u) = b_x(t,x(t),u(t)) \text{ and } g_z(t,u) = g_z(t,x(t),y(t),z(t),\tilde{z}(t),u(t)).$$

For any $v(\cdot) \in \mathscr{U}_{ad}$, it is easy to see that under (H1.6), (4.4) and (4.5) admit a unique solution, respectively.

Lemma 4.3. *Under (H1.6), it follows that*

$$\mathbb{E}|x^1(t)|^8 < \infty, \quad \mathbb{E}|Z^1(t)|^4 < \infty. \tag{4.6}$$

Let

$$\phi^\varepsilon(t) = \frac{\phi^{u+\varepsilon v}(t) - \phi(t)}{\varepsilon} - \phi^1(t) \text{ with } \phi = x, y, z, \tilde{z}, Z. \tag{4.7}$$

Note that ϕ^ε defined in (4.7) is for $\varepsilon \in [0,1)$, and it should not be confused with ϕ^1 defined in (4.5).

Lemma 4.4. *If (H1.6) holds, then*

$$
\begin{aligned}
&\lim_{\varepsilon\to 0} \sup_{0\le t\le T} \mathbb{E}|x^\varepsilon(t)|^4 = 0, \quad \lim_{\varepsilon\to 0} \sup_{0\le t\le T} \mathbb{E}|Z^\varepsilon(t)|^2 = 0, \\
&\lim_{\varepsilon\to 0} \mathbb{E}\int_0^T |z^\varepsilon(t)|^2 dt = 0, \quad \lim_{\varepsilon\to 0} \mathbb{E}\int_0^T |\tilde{z}^\varepsilon(t)|^2 dt = 0, \\
&\lim_{\varepsilon\to 0} \sup_{0\le t\le T} \mathbb{E}|y^\varepsilon(t)|^2 = 0.
\end{aligned}
$$

Proof. It follows from (1.15) and (4.5) that

$$dx^\varepsilon(t) = b^\varepsilon(t)dt + \sigma^\varepsilon(t)dW(t) + \tilde{\sigma}^\varepsilon(t)dY(t),$$

where

$$
\begin{aligned}
b^\varepsilon(t) &= \left(\frac{b(t,u+\varepsilon v)-b(t,u)}{\varepsilon} - b_x(t,u)x^1(t) - b_v(t,u)v(t)\right) \\
&\quad - \left(\frac{\tilde\sigma(t,u+\varepsilon v)-\tilde\sigma(t,u)}{\varepsilon} - \tilde\sigma_x(t,u)x^1(t) - \tilde\sigma_v(t,u)v(t)\right) h(t,x) \\
&\quad - \left(\frac{h(t,x^{u+\varepsilon v})-h(t,x)}{\varepsilon} - h_x(t,x)x^1(t)\right)\tilde\sigma(t,u) \\
&\quad - \frac{\tilde\sigma(t,u+\varepsilon v)-\tilde\sigma(t,u)}{\varepsilon}\left(h(t,x^{u+\varepsilon v})-h(t,x)\right),
\end{aligned}
$$

$$
\sigma^\varepsilon(t) = \frac{\sigma(t,u+\varepsilon v)-\sigma(t,u)}{\varepsilon} - \sigma_x(t,u)x^1(t) - \sigma_v(t,u)v(t),
$$

and

$$
\tilde\sigma^\varepsilon(t) = \frac{\tilde\sigma(t,u+\varepsilon v)-\tilde\sigma(t,u)}{\varepsilon} - \tilde\sigma_x(t,u)x^1(t) - \tilde\sigma_v(t,u)v(t).
$$

Denote

$$
\Theta = (t, x+\varepsilon\lambda(x^\varepsilon + x^1), u+\varepsilon\lambda v) \text{ and } \Xi = (t, x+\varepsilon\lambda(x^\varepsilon + x^1)).
$$

It is easy to show that

$$
\begin{aligned}
\sigma^\varepsilon(t) &= x^\varepsilon(t)\int_0^1 \sigma_x(\Theta)d\lambda + x^1(t)\left(\int_0^1 \sigma_x(\Theta)d\lambda - \sigma_x(t,u)\right) \\
&\quad + v(t)\left(\int_0^1 \sigma_v(\Theta)d\lambda - \sigma_v(t,u)\right).
\end{aligned}
$$

Denote by $\gamma^\varepsilon(t)$ the maximum of

$$
\left|\phi_X(t,x+\varepsilon\lambda(x^\varepsilon+x^1),u+\varepsilon\lambda v) - \phi_X(t,x,u)\right| \tag{4.8}
$$

for ϕ and X runs over σ, $\tilde\sigma$, b, h and x, v, respectively. Then, by (H1.6) and Lemmas 4.2 and 4.3, we have

$$
|\sigma^\varepsilon(t)| \le K\left(|x^\varepsilon(t)| + \left(|x^1(t)| + |v(t)|\right)\gamma^\varepsilon(t)\right). \tag{4.9}
$$

Similarly, we can prove that

$$
|\tilde\sigma^\varepsilon(t)| \le K\left(|x^\varepsilon(t)| + \left(|x^1(t)| + |v(t)|\right)\gamma^\varepsilon(t)\right),
$$

and

$$
|b^\varepsilon(t)| \le K\left(|x^\varepsilon(t)| + \left(|x^1(t)| + |v(t)|\right)(\gamma^\varepsilon(t)\vee\varepsilon)\right).
$$

According to Hölder's inequality and the BDG inequality, we derive

$$\mathbb{E}|x^\varepsilon(t)|^4 \le C\mathbb{E}\int_0^t |x^\varepsilon(s)|^4 ds$$
$$+C\int_0^T \left(\mathbb{E}|x^1(t)|^8+\mathbb{E}|v(t)|^8\right)^{1/2}\left(\mathbb{E}\left(\gamma^\varepsilon(t)\vee\varepsilon\right)^8\right)^{1/2} dt.$$

Note that $\mathbb{E}\left(\gamma^\varepsilon(t)^8\right)\to 0$. By Gronwall's inequality, we obtain the first limit. The second can be proved similarly.

To prove the other limit, we note that

$$-dy^\varepsilon(t) = g^\varepsilon(t)dt - z^\varepsilon(t)dW(t) - \bar{z}^\varepsilon(t)dY(t),$$

where

$$g^\varepsilon(t) = \varepsilon^{-1}\left(g(t,x^{u+\varepsilon v},y^{u+\varepsilon v},z^{u+\varepsilon v},\bar{z}^{u+\varepsilon v},u+\varepsilon v) - g(t,x,y,z,\bar{z},u)\right)$$
$$-\left(g_x(t)x^1(t)+g_y(t)y^1(t)+g_z(t)z^1(t)+g_{\bar{z}}(t)\bar{z}^1(t)+g_v(t)v(t)\right).$$

Applying Itô's formula, we get

$$d|y^\varepsilon(t)|^2 = \left(-2g^\varepsilon(t)y^\varepsilon(t)+|z^\varepsilon(t)|^2+|\bar{z}^\varepsilon(t)|^2\right)dt$$
$$+2y^\varepsilon(t)z^\varepsilon(t)dW(t)+2y^\varepsilon(t)\bar{z}^\varepsilon(t)dY(t).$$

Taking integral and then expectation, we have

$$\begin{aligned}\mathbb{E}|y^\varepsilon(t)|^2-\mathbb{E}|y^\varepsilon(T)|^2 &= \mathbb{E}\int_t^T 2g^\varepsilon(s)y^\varepsilon(s)ds - \mathbb{E}\int_t^T\left(|z^\varepsilon(s)|^2+|\bar{z}^\varepsilon(s)|^2\right)ds\\ &\le \delta\mathbb{E}\int_t^T |g^\varepsilon(s)|^2ds+\delta^{-1}\mathbb{E}\int_t^T|y^\varepsilon(s)|^2ds\\ &\quad -\mathbb{E}\int_t^T\left(|z^\varepsilon(s)|^2+|\bar{z}^\varepsilon(s)|^2\right)ds, \end{aligned} \tag{4.10}$$

where $\delta>0$ is an arbitrary constant.

Similar to (4.9), we can prove that $\mathbb{E}|y^\varepsilon(T)|^2\to 0$ and

$$\mathbb{E}|g^\varepsilon(s)|^2 \le K\mathbb{E}\left(|x^\varepsilon(s)|^2+|y^\varepsilon(s)|^2+|z^\varepsilon(s)|^2+|\bar{z}^\varepsilon(s)|^2\right)$$
$$+K\mathbb{E}\left(\left(|x^1(s)|^2+|y^1(s)|^2+|z^1(s)|^2+|\bar{z}^1(s)|^2+|v(s)|^2\right)\left(\tilde{\gamma}^\varepsilon(s)\vee\varepsilon\right)\right),$$

where K is a constant which may depend on x^1 etc., and $\tilde{\gamma}^\varepsilon$ is defined as (4.8) with $\phi=g$ and X runs over x, y, z, $\bar{z}$, v.

Note that $\tilde{\gamma}^\varepsilon$ is bounded and convergent to 0, by the dominated convergent theorem, we have

$$\int_0^T\mathbb{E}\left(\left(|x^1(s)|^2+|y^1(s)|^2+|z^1(s)|^2+|\bar{z}^1(s)|^2+|v(s)|^2\right)\left(\tilde{\gamma}^\varepsilon(s)\vee\varepsilon\right)\right)ds\to 0.$$

Taking δ small enough such that $\delta K<1$ in (4.10), it then follows from Gronwall's inequality that the last three identities of the lemma hold. □

4.1.2 Stochastic Maximum Principle

The following assumption and adjoint equations will be needed in deriving the stochastic maximum principle.

(H4.1) *(i) For any t, τ such that $t+\tau \in [0,T]$, and bounded $\mathscr{F}_t^Y$-measurable random variable ν, we formulate the control process $v(s) \in U$, with*

$$v(s) = \nu I_{[t,t+\tau)}(s), \quad s \in [0,T],$$

where $I_{[t,t+\tau)}(s)$ is the indicator function on the set $[t,t+\tau]$.

(ii) For any $v(s) \in \mathscr{F}_s^Y$ with $v(s)$ bounded, $s \in [0,T]$, there is an $\varepsilon > 0$ such that $u(\cdot)+\varepsilon v(\cdot) \in \mathscr{U}_{ad}$ for $\varepsilon \in (-1,1)$.

We formulate the adjoint equations

$$\begin{cases} dp(t) = [g_y(t,u)p(t) - l_y(t,u)]\,dt \\ \qquad + [g_z(t,u)p(t) - l_z(t,u)]\,dW(t) \\ \qquad + [(g_{\tilde{z}}(t,u) - h(t,x(t)))\,p(t) - l_{\tilde{z}}(t,u)]\,d\tilde{W}(t), \\ -dq(t) = \{[b_x(t,u) - \tilde{\sigma}(t,u)h_x(t,x(t))]\,q(t) \\ \qquad + \sigma_x(t,u)k(t) + \tilde{\sigma}_x(t,u)\tilde{k}(t) + h_x(t,x(t))\tilde{Q}(t) \\ \qquad - g_x(t,u)p(t) + l_x(t,u)\}\,dt \\ \qquad - k(t)dW(t) - \tilde{k}(t)d\tilde{W}(t), \\ p(0) = -\gamma_y(y(0)), \quad q(1) = -f_x(x(T))p(T) + \phi_x(x(T)), \end{cases} \tag{4.11}$$

and

$$\begin{cases} -dP(t) = l(t,u)dt - Q(t)dW(t) - \tilde{Q}(t)d\tilde{W}(t), \\ P(T) = \phi(x(T)). \end{cases} \tag{4.12}$$

Hereinafter we adopt the notation $\tilde{W}(\cdot) = \tilde{W}^u(\cdot)$. Note that the appearance of the driving Brownian motion $\tilde{W}^v(\cdot)$ in (4.1) makes adjoint equations (4.12) and (4.11) dramatically different from the classical FBSDEs. Moreover, (1.25) is used to treat the terms induced by partially observable information, which is unnecessary in the cases of Peng [66], Øksendal and Sulem [61], Wu [100], and Yong [107].

We now state the first maximum principle for optimal control of Problem B.

Theorem 4.1. *Let (H1.6), (H1.7), and (H4.1) hold. Assume that $u(\cdot)$ is a local minimum for $J(v(\cdot))$, in the sense that for all process $v(\cdot)$ such that $v(\cdot)+u(\cdot) \in \mathscr{U}_{ad}$,*

$$\mathscr{J}(\varepsilon) = J(u(\cdot)+\varepsilon v(\cdot)), \quad \varepsilon \in [0,1)$$

attains its minimum at $\varepsilon = 0$. Suppose that for any $v(\cdot) \in \mathscr{U}_{ad}$, the functions ϕ, $\phi_x \in \mathscr{L}_{\mathbb{F}}^2(\Omega;\mathbb{R})$, l, l_x, l_y, l_z, $l_{\tilde{z}}$, $l_v \in \mathscr{L}_{\mathbb{F}}^2(0,T;\mathbb{R})$. Furthermore, suppose that

(1.25) and (1.26) admit unique solutions $(P(\cdot),Q(\cdot),\tilde{Q}(\cdot)) \in \mathscr{L}^2_{\mathscr{F}}(0,T;\mathbb{R}^3)$ *and* $(p(\cdot),q(\cdot),k(\cdot),\tilde{k}(\cdot)) \in \mathscr{L}^2_{\mathbb{F}}(0,T;\mathbb{R}^4)$*, respectively. Then for any* $v \in U$ *we have*

$$\mathbb{E}^u\left[H_v(t,x(t),y(t),z(t),\tilde{z}(t),u(t);p(t),q(t),k(t),\tilde{k}(t),\tilde{Q}(t))(v-u(t))|\mathscr{F}_t^Y\right] \geq 0,$$

where the Hamiltonian function $H : [0,T] \times \mathbb{R}^4 \times U \times \mathbb{R}^5 \to \mathbb{R}$ *is defined by*

$$\begin{aligned} H(t,x,y,z,\tilde{z},v;p,q,k,\tilde{k},\tilde{Q}) &= b(t,x,v)q + \sigma(t,x,v)k + \tilde{\sigma}(t,x,v)\tilde{k} + h(t,x)\tilde{Q} \\ &\quad - \Big(g(t,x,y,z,\tilde{z},v) - h(t,x)\tilde{z}\Big)p + l(t,x,y,z,\tilde{z},v). \end{aligned} \tag{4.13}$$

Proof. Note that

$$\begin{aligned} 0 &\leq \frac{d}{d\varepsilon}\mathscr{J}(\varepsilon)\Big|_{\varepsilon=0} \\ &= \lim_{\varepsilon\to 0}\frac{J(u(\cdot)+\varepsilon v(\cdot)) - J(u(\cdot))}{\varepsilon} \\ &= \lim_{\varepsilon\to 0}\frac{1}{\varepsilon}\mathbb{E}\Bigg\{\int_0^T \left[\left(Z^{u+\varepsilon v}(t) - Z(t)\right)l(t,u)\right. \\ &\qquad \left. + Z^{u+\varepsilon v}(t)\left(l(t,u(t)+\varepsilon v(t)) - l(t,u)\right)\right]dt \\ &\qquad + \left(Z^{u+\varepsilon v}(T) - Z(T)\right)\phi(x(T)) + Z^{u+\varepsilon v}(T)\left(\phi(x^{u+\varepsilon v}(T)) - \phi(x(T))\right) \\ &\qquad + \gamma(y^{u+\varepsilon v}(0)) - \gamma(y(0))\Bigg\}. \end{aligned} \tag{4.14}$$

To deal with $Z(\cdot)$, let $\Gamma(\cdot) = Z^1(\cdot)Z^{-1}(\cdot)$. Making use of (1.16) and (4.4), by Itô's formula, we get

$$\begin{cases} d\Gamma(t) = h_x(t,x(t))x^1(t)\left(dY(t) - h(t,x(t))dt\right) \\ \qquad\quad = h_x(t,x(t))x^1(t)d\tilde{W}(t), \\ \Gamma(0) = 0. \end{cases} \tag{4.15}$$

Applying Itô's formula to $P(\cdot)\Gamma(\cdot)$, $p(\cdot)y^1(\cdot)$, and $q(\cdot)x^1(\cdot)$, respectively, we derive

$$\begin{aligned} &\mathbb{E}^u\left[\Gamma(T)\phi(x(T)) + \int_0^T \Gamma(t)l(t,u)dt\right] \\ &= \mathbb{E}^u\int_0^T \tilde{Q}(t)h_x(t,x(t))x^1(t)dt, \end{aligned} \tag{4.16}$$

$$\begin{aligned} &\mathbb{E}^u\left[p(T)f_x(x(T))x^1(T) + \gamma_y(y(0))y^1(0)\right] \\ &= -\mathbb{E}^u\int_0^T \left[l_y(t,u)y^1(t) + l_z(t,u)z^1(t) + l_{\tilde{z}}(t,u)\tilde{z}^1(t)\right]dt \\ &\quad - \mathbb{E}^u\int_0^T \left[g_v(t,u)v(t) + g_x(t,u)x^1(t)\right]p(t)dt \end{aligned} \tag{4.17}$$

and

$$
\begin{aligned}
&\mathbb{E}^u\left[\phi_x(x(T))x^1(T) - p(T)f_x(x(T))x^1(T)\right] \\
&= \mathbb{E}^u\int_0^T g_x(t,u)x^1(t)p(t)dt \\
&\quad - \mathbb{E}^u\int_0^T \left[l_x(t,u) + \tilde{Q}(t)h_x(t,x)\right]x^1(t)dt \\
&\quad + \mathbb{E}^u\int_0^T \left[b_v(t,u)q(t) + \sigma_v(t,u)k(t) + \tilde{\sigma}_v(t,u)\tilde{k}(t)\right]v(t)dt.
\end{aligned}
\tag{4.18}
$$

By Lemmas 4.2 and 4.4, we may continue (4.14) with

$$
\begin{aligned}
0 \leq\ & \mathbb{E}^u\left[\phi_x(x(T))x^1(T) + \gamma_y(y(0))y^1(0)\right] \\
&+ \mathbb{E}^u\left[\phi(x(T))\Gamma(T) + \int_0^T \Gamma(t)l(t,u)dt\right] \\
&+ \mathbb{E}^u\int_0^T \left[l_x(t,u)x^1(t) + l_y(t,u)y^1(t)\right]dt \\
&+ \mathbb{E}^u\int_0^T \left[l_z(t,u)z^1(t) + l_{\tilde{z}}(t,u)\tilde{z}^1(t)\right]dt \\
&+ \mathbb{E}^u\int_0^T l_v(t,u)v(t)dt.
\end{aligned}
\tag{4.19}
$$

Substituting (4.16), (4.17), and (4.18) into (4.19) and recalling Condition (H4.1), we have

$$
\begin{aligned}
0 \leq\ & \mathbb{E}^u\int_0^T \left[b_v(t,u)q(t) + \sigma_v(t,u)k(t) + \tilde{\sigma}(t,u)\tilde{k}(t)\right]v(t)dt \\
&+ \mathbb{E}^u\int_0^T \left[l_v(t,u) - g_v(t,u)p(t)\right]v(t)dt \\
=\ & \mathbb{E}^u\int_t^{t+\tau} vH_v(s,x,y,z,\tilde{z},u;p,q,k,\tilde{k},\tilde{Q})ds.
\end{aligned}
\tag{4.20}
$$

Differentiating with respect to τ, we get

$$
\mathbb{E}^u\left[vH_v(t,x,y,z,\tilde{z},u;p,q,k,\tilde{k},\tilde{Q})|\mathscr{F}_t^Y\right] \geq 0.
$$

The proof is then completed. □

4.2 A Malliavin Derivative Method

We now state the second maximum principle for optimal control of Problem B.

Theorem 4.2. *Let (H1.6), (H1.7), and (H4.1) hold. Assume that $u(\cdot)$ is a local minimum for $J(v(\cdot))$, in the sense that for all processes $v(\cdot)$ with $u(\cdot)+v(\cdot) \in \mathscr{U}_{ad}$,*

$$\mathscr{J}(\varepsilon) = J(u(\cdot) + \varepsilon v(\cdot)), \quad \varepsilon \in [0,1]$$

attains its minimum at $\varepsilon = 0$. *Suppose that (1.27) admits the unique solution* $\bar{p}(\cdot) \in \mathscr{L}^2_{\mathscr{F}}(0,T;\mathbb{D}_{1,2})$. *Assume that* ϕ, $\phi_x \in \mathbb{D}_{1,2}$, l, l_x, *and* $\Psi(t,s)$ *are in* $\mathbb{L}_{1,2}(\mathbb{R})$ *for all* $0 \le t \le s \le T$. *Then for any* $v \in U$ *we have*

$$\mathbb{E}^u \left[\bar{H}_v(t,x(t),y(t),z(t),\tilde{z}(t),u(t);\bar{p}(t),\bar{q}(t),\bar{k}(t),\bar{\bar{k}}(t))(v-u(t)) \Big| \mathscr{F}_t^Y \right] \ge 0,$$

where $\bar{H}_v$ *is defined by*

$$\begin{aligned}\bar{H}_v(t,x,y,z,\bar{z},v;\bar{p},\bar{q},\bar{k},\bar{\bar{k}}) &= b_v(t,x,v)\bar{q} + \sigma_v(t,x,v)\bar{k} + \tilde{\sigma}_v(t,x,v)\bar{\bar{k}} \\ &\quad - g_v(t,x,y,z,\tilde{z},v)\bar{p} + l_v(t,x,y,z,\bar{z},v).\end{aligned}$$

Proof. If $u(\cdot)$ is a local minimum for $J(v(\cdot))$, then

$$\begin{aligned}0 &\le \frac{d}{d\varepsilon}\mathscr{J}(\varepsilon)\Big|_{\varepsilon=0} \\ &= \mathbb{E}^u \left(\phi(x(T))\Gamma(T) + \int_0^T \Gamma(t)l(t,u)dt \right) \\ &\quad + \mathbb{E}^u \left[(\phi_x(x(T)) - \bar{p}(T)f_x(x(T)))x^1(T) \right] \\ &\quad + \mathbb{E}^u \left(\bar{p}(T)f_x(x(T))x^1(T) + \gamma_y(y(0))y^1(0) \right) \\ &\quad + \mathbb{E}^u \int_0^T \left(l_x(t,u)x^1(t) + l_y(t,u)y^1(t) \right) dt \\ &\quad + \mathbb{E}^u \int_0^T \left(l_z(t,u)z^1(t) + l_{\tilde{z}}(t,u)\tilde{z}^1(t) \right) dt \\ &\quad + \mathbb{E}^u \int_0^T l_v(t,u)v(t)dt.\end{aligned} \tag{4.21}$$

According to (4.15), Lemmas A.7 and A.8, we have

$$\begin{aligned}\mathbb{E}^u(\phi(x(T))\Gamma(T)) &= \mathbb{E}^u \left(\phi(x(T)) \int_0^T h_x(t,x)x^1(t)d\tilde{W}(t) \right) \\ &= \mathbb{E}^u \int_0^T h_x(t,x)x^1(t)D_t^{(\tilde{W})}\phi(x(T))dt\end{aligned} \tag{4.22}$$

and

$$\begin{aligned}\mathbb{E}^u \int_0^T \Gamma(t)l(t,u)dt &= \mathbb{E}^u \int_0^T l(t,u) \int_0^t h_x(s,x)x^1(s)d\tilde{W}(s)dt \\ &= \mathbb{E}^u \int_0^T \int_0^t h_x(t,x)x^1(t)D_s^{(\tilde{W})}l(t,u)dsdt \\ &= \mathbb{E}^u \int_0^T \left(\int_t^T D_t^{(\tilde{W})}l(s,u)ds \right) h_x(t,x)x^1(t)dt.\end{aligned} \tag{4.23}$$

Note that, in deriving the last line in (4.23), we used Fubini's theorem. It then follows from (4.22) and (4.23) that

$$
\begin{aligned}
&\mathbb{E}^u\left(\phi(x(T))\Gamma(T)+\int_0^T \Gamma(t)l(t,u)dt\right)\\
&=\mathbb{E}^u\int_0^T\left(D_t^{(\tilde{W})}\phi(x(T))+\int_t^T D_t^{(\tilde{W})}l(s,u)ds\right)h_x(t,x)x^1(t)dt \qquad (4.24)\\
&=\mathbb{E}^u\int_0^T h_x(t,x)x^1(t)D_t^{(\tilde{W})}\Pi(t)dt.
\end{aligned}
$$

Similarly,

$$
\begin{aligned}
&\mathbb{E}^u\left[(\phi_x(x(T))-\bar{p}(T)f_x(x(T)))x^1(T)\right] \qquad (4.25)\\
=\ &\mathbb{E}^u\Big\{(\phi_x(x(T))-\bar{p}(T)f_x(x(T)))\\
&\times\left[\int_0^T((b_x(t,u)-\tilde{\sigma}(t,u)h_x(t,x))x^1(t)+b_v(t,u)v(t))\,dt\right.\\
&+\int_0^T(\sigma_x(t,u)x^1(t)+\sigma_v(t,u)v(t))\,dW(t)\\
&\left.+\int_0^T(\tilde{\sigma}_x(t,u)x^1(t)+\tilde{\sigma}_v(t,u)v(t))\,d\tilde{W}(t)\right]\Big\}\\
=\ &\mathbb{E}^u\int_0^T(\phi_x(x(T))-\bar{p}(T)f_x(x(T)))\\
&\times\Big\{(b_x(t,u)-\tilde{\sigma}(t,u)h_x(t,x))x^1(t)+b_v(t,u)v(t)\\
&+(\sigma_x(t,u)x^1(t)+\sigma_v(t,u)v(t))\,D_t^{(W)}(\phi_x(x(T))-\bar{p}(T)f_x(x(T)))\\
&+(\tilde{\sigma}_x(t,u)x^1(t)+\tilde{\sigma}_v(t,u)v(t))\,D_t^{(\tilde{W})}(\phi_x(x(T))-\bar{p}(T)f_x(x(T)))\Big\}\,dt.
\end{aligned}
$$

By (4.5), and Lemmas A.7 and A.8, we have

$$
\begin{aligned}
&\mathbb{E}^u\int_0^T l_x(t,u)x^1(t)dt\\
&=\mathbb{E}^u\int_0^T\int_0^t\Big\{l_x(t,u)\left[(b_x(s,u)-\tilde{\sigma}(s,u)h_x(s,x))x^1(s)+b_v(s,u)v(s)\right]\\
&+(\sigma_x(s,u)x^1(s)+\sigma_v(s,u)v(s))\,D_s^{(W)}l_x(t,u)\\
&+(\tilde{\sigma}_x(s,u)x^1(s)+\tilde{\sigma}_v(s,u)v(s))\,D_s^{(\tilde{W})}l_x(t,u)\Big\}\,dsdt. \qquad (4.26)
\end{aligned}
$$

Simple calculations from (4.26) then yield that

$$
\begin{aligned}
&\mathbb{E}^u \int_0^T l_x(t,u)x^1(t)dt \\
&= \mathbb{E}^u \int_0^T \bigg\{ \int_t^T l_x(s,u)ds \left[(b_x(t,u) - \tilde{\sigma}(t,u)h_x(t,x))x^1(t) + b_v(t,u)v(t) \right] \\
&\quad + \left(\sigma_x(t,u)x^1(t) + \sigma_v(t,u)v(t) \right) \int_t^T D_t^{(W)} l_x(s,u)ds \\
&\quad + \left(\tilde{\sigma}_x(t,u)x^1(t) + \tilde{\sigma}_v(t,u)v(t) \right) \int_t^T D_t^{(\tilde{W})} l_x(s,u)ds \bigg\} dt. \qquad (4.27)
\end{aligned}
$$

By (4.25) and (4.27) we then have

$$
\begin{aligned}
&\mathbb{E}^u \left[(\phi_x(x(T)) - \bar{p}(T)f_x(x(T)))x^1(T) + \int_0^T l_x(t,u)x^1(t)dt \right] \\
&= \mathbb{E}^u \int_0^T \Big\{ \Sigma(t) \left[(b_x(t,u) - \tilde{\sigma}(t,u)h_x(t,x))x^1(t) + b_v(t,u)v(t) \right] \\
&\quad + \left(\sigma_x(t,u)x^1(t) + \sigma_v(t,u)v(t) \right) D_t^{(W)} \Sigma(t) \\
&\quad + \left(\tilde{\sigma}_x(t,u)x^1(t) + \tilde{\sigma}_v(t,u)v(t) \right) D_t^{(\tilde{W})} \Sigma(t) \Big\} dt. \qquad (4.28)
\end{aligned}
$$

Applying Itô's formula to $\bar{p}(\cdot)y^1(\cdot)$, we derive

$$
\begin{aligned}
&\mathbb{E}^u \left[\bar{p}(T)f_x(x(T))x^1(T) + \gamma_y(y(0))y^1(0) \right] \\
&= -\mathbb{E}^u \int_0^T \left[l_y(t,u)y^1(t) + l_z(t,u)z^1(t) + l_{\tilde{z}}(t,u)\tilde{z}^1(t) \right] dt \qquad (4.29) \\
&\quad - \mathbb{E}^u \int_0^T \left[g_v(t,u)v(t) + g_x(t,u)x^1(t) \right] \bar{p}(t)dt.
\end{aligned}
$$

Inserting (4.24), (4.28), and (4.29) into (4.21), we have

$$
\begin{aligned}
0 &\leq \frac{d}{d\varepsilon} \mathscr{J}(\varepsilon) \Big|_{\varepsilon=0} \\
&= \mathbb{E}^u \int_0^T \Big[\Sigma(t) (b_x(t,u) - \tilde{\sigma}(t,u)h_x(t,x)) + \sigma_x(t,u)D_t^{(W)} \Sigma(t) \\
&\quad + \tilde{\sigma}_x(t,u)D_t^{(\tilde{W})} \Sigma(t) + h_x(t,x)D_t^{(\tilde{W})} \Pi(t) - g_x(t,x,y,z,\tilde{z},u)\bar{p}(t) \Big] x^1(t)dt \qquad (4.30) \\
&\quad + \mathbb{E}^u \int_0^T \Big[\Sigma(t)b_v(t,u) + \sigma_v(t,u)D_t^{(W)} \Sigma(t) + \tilde{\sigma}_v(t,u)D_t^{(\tilde{W})} \Sigma(t) \\
&\quad + l_v(t,u) - g_v(t,u)\bar{p}(t) \Big] v(t)dt.
\end{aligned}
$$

Since (4.30) holds for any admissible control $v(\cdot)$, hereafter we take

$$
v(s) = vI_{(t,t+\tau]}(s),
$$

where $v = v(\omega)$ is a bounded $\mathscr{F}_t^Y$-measurable random variable, $0 \le t \le t+\tau \le T$. In this situation, it is easy to see from (4.5) that

$$x^1(s) = 0, \quad \text{for } 0 \le s \le t. \tag{4.31}$$

Then (4.30) can be written as

$$0 \le \mathscr{I}_1(\tau) + \mathscr{I}_2(\tau) \tag{4.32}$$

with

$$\begin{aligned}\mathscr{I}_1(\tau) = \mathbb{E}^u \int_t^T & \Big[\Sigma(s)\left(b_x(s,u) - \tilde{\sigma}(s,u)h_x(s,x)\right) + \sigma_x(s,u)D_s^{(W)}\Sigma(s) \\ & + \tilde{\sigma}_x(s,u)D_s^{(\tilde{W})}\Sigma(s) + h_x(s,x)D_s^{(\tilde{W})}\Pi(s) \\ & - g_x(s,u)\bar{p}(s)\big]x^1(s)ds\end{aligned} \tag{4.33}$$

and

$$\begin{aligned}\mathscr{I}_2(\tau) = \mathbb{E}^u \int_t^{t+\tau} v & \Big[\Sigma(s)b_v(s,u) + \sigma_v(s,u)D_s^{(W)}\Sigma(s) + \tilde{\sigma}_v(s,u)D_s^{(\tilde{W})}\Sigma(s) \\ & + l_v(s,u) - g_v(s,u)\bar{p}(s)\Big]ds.\end{aligned} \tag{4.34}$$

Note that with the special control $v(s) = vI_{(t,t+\tau]}(s)$, we arrive at

$$\begin{aligned}dx^1(s) = x^1(s)\Big\{ & [b_x(s,u) - \tilde{\sigma}(s,u)h_x(s,x)]ds \\ & + \sigma_x(s,u)dW(s) + \tilde{\sigma}_x(s,u)d\tilde{W}(s)\Big\}, \quad \text{for } s \ge t+\tau.\end{aligned}$$

Solving the above equation, we get

$$x^1(s) = x^1(t+\tau)\Phi(t+\tau,s),$$

where

$$\begin{aligned}x^1(t+\tau) = v & \int_t^{t+\tau} \left(b_v(r,u)dr + \sigma_v(r,u)dW(r) + \tilde{\sigma}_v(r,u)d\tilde{W}(r)\right) \\ & + \int_t^{t+\tau} x^1(r)\left[\left(b_x(r,u) - \tilde{\sigma}(r,u)h_x(r,x)\right)dr\right. \\ & \left. + \sigma_x(r,u)dW(r) + \tilde{\sigma}_x(r,u)d\tilde{W}(r)\right].\end{aligned}$$

Then

$$\begin{aligned}\frac{d}{d\tau}\mathscr{J}_1(\tau)\Big|_{\tau=0} &= \frac{d}{d\tau}\mathbb{E}^u\left[\int_{t+\tau}^T H_x(s)x^1(t+\tau)\Phi(t+\tau,s)ds\right]_{\tau=0}\\ &= \int_t^T \frac{d}{d\tau}\mathbb{E}^u\left[H_x(s)x^1(t+\tau)\Phi(t+\tau,s)\right]_{\tau=0}ds\\ &= \int_t^T \frac{d}{d\tau}\mathbb{E}^u\left[x^1(t+\tau)\Psi(t,s)\right]_{\tau=0}ds.\\ &= \mathscr{J}_{11}+\mathscr{J}_{12},\end{aligned}$$

where

$$\begin{aligned}\mathscr{J}_{11} = \int_t^T \frac{d}{d\tau}\mathbb{E}^u\Big\{\Psi(t,s)\int_t^{t+\tau} x^1(r)\left[(b_x(r,u)-\tilde{\sigma}(r,u)h_x(r,x))dr\right.\\ \left.+\sigma_x(r,u)dW(r)+\tilde{\sigma}_x(r,u)d\tilde{W}(r)\right]\Big\}_{\tau=0} ds\end{aligned} \tag{4.35}$$

and

$$\begin{aligned}\mathscr{J}_{12} = \int_t^T \frac{d}{d\tau}\mathbb{E}^u\Big\{v\Psi(t,s)\int_t^{t+\tau}\left[b_v(r,u)dr\right.\\ \left.+\sigma_v(r,u)dW(r)+\tilde{\sigma}_v(r,u)d\tilde{W}(r)\right]\Big\}_{\tau=0} ds.\end{aligned} \tag{4.36}$$

According to (4.31), Lemmas A.7 and A.8, it is not difficult to derive that

$$\mathscr{J}_{11}=0$$

and

$$\begin{aligned}\mathscr{J}_{12} = \mathbb{E}^u\int_t^T v\Big(\Psi(t,s)b_v(t,u)+\sigma_v(t,u)D_t^{(W)}\Psi(t,s)\\ +\tilde{\sigma}_v(t,u)D_t^{(\tilde{W})}\Psi(t,s)\Big)ds.\end{aligned} \tag{4.37}$$

Similarly,

$$\begin{aligned}\frac{d}{d\tau}\mathscr{J}_2(\tau)\Big|_{\tau=0} = \mathbb{E}^u\Big\{v\Big[\Sigma(t)b_v(t,u)+\sigma_v(t,u)D_t^{(W)}\Sigma(t)\\ +\tilde{\sigma}_v(t,u)D_t^{(\tilde{W})}\Sigma(t)+l_v(t,u)-g_v(t,u)\bar{p}(t)\Big]\Big\}.\end{aligned} \tag{4.38}$$

From (4.21), (4.37), and (4.38), we get

$$\begin{aligned}0 &\le \frac{d}{d\varepsilon}\mathscr{J}(\varepsilon)\Big|_{\varepsilon=0}\\ &= \mathbb{E}^u\Big\{v\Big[b_v(t,u)\bar{q}(t)+\sigma_v(t,u)\bar{k}(t)+\tilde{\sigma}_v(t,u)\bar{\bar{k}}(t)\\ &\quad +l_v(t,u)-g_v(t,u)\bar{p}(t)\Big]\Big\}.\end{aligned}$$

The proof is then completed. □

4.3 A Recursive Utility Optimization Problem

This section focuses on illustrating Theorem 4.2 within the framework of recursive utility. For convenience, we let $\tilde{C}(t) = 0$ in (1.7), $0 \le t \le T$.

The aim of the policymaker is to find a control strategy $u(\cdot) \in \mathscr{U}_{ad}$ so that

$$J(u(\cdot)) = \min_{v(\cdot)\in\mathscr{U}_{ad}} \mathbb{E}^v \left[\frac{1}{2}\int_0^T (v(t)-M(t))^2\,dt - y^v(0)\right] \tag{4.39}$$

subject to (1.7), (1.8) and Definition 1.2, where $M(\cdot)$ is a pre-set target, and $y^v(\cdot)$ is a generalized recursive utility resulting from x and v. In the sense of El Karoui et al. [19], $y^v(\cdot)$ can be regarded as the solution of

$$\begin{cases} -dy^v(t) = g(t,x^v(t),y^v(t),z^v(t),\tilde{z}^v(t))dt - z^v(t)dW(t) - \tilde{z}^v(t)dY(t), \\ y^v(T) = f(x^v(T)), \end{cases}$$

where f and g satisfy (H1.6). The example captures the scenario where the policymaker has two objectives: on one hand, the concern of the policymaker is to prevent the control strategy $v(\cdot)$ from large deviations so as to stabilize the related economic scheme, on the other hand, he/she would like to optimize the recursive utility. Note that utility functional (4.39) is inspired by Shi and Wu [75], where an optimization problem with complete information was studied.

With this setup, it is easy to see from (1.7) and (1.8) that

$$\begin{aligned} &b(t,x,v) = A(t)x + B(t)v,\ \sigma(t,x,v) = C(t)v + D(t),\\ &\tilde{\sigma}(t,x,v) = \tilde{D}(t),\ h(t,x) = \tfrac{1}{\beta}\alpha(t,x) - \tfrac{1}{2}\beta. \end{aligned}$$

The new adjoint processes are written as

$$\begin{aligned} &\bar{q}(t) = -f_x(x(T))\bar{p}(T) + \int_t^T H_x(s)\Phi(t,s)ds,\\ &\bar{k}(t) = D_t^{(W)}\bar{q}(t), \quad \bar{\bar{k}}(t) = D_t^{(\tilde{W})}\bar{q}(t) \end{aligned} \tag{4.40}$$

with

$$\begin{cases} d\bar{p}(t) = \bar{p}(t)\Big[g_y(t,x(t),y(t),z(t),\tilde{z}(t))dt + g_z(t,x(t),y(t),z(t),\tilde{z}(t))dW(t) \\ \qquad + \Big(g_{\tilde{z}}(t,x(t),y(t),z(t),\tilde{z}(t)) - \dfrac{1}{\beta}\alpha(t,x(t)) + \dfrac{1}{2}\beta\Big)d\tilde{W}(t)\Big], \\ \bar{p}(0) = 1, \end{cases}$$

$$\begin{aligned} H_x(t) = &-f_x(x(T))\bar{p}(T)\left[A(t) - \frac{1}{\beta}\tilde{D}(t)\alpha_x(t,x(t))\right] + \frac{1}{\beta}\alpha_x(t,x(t))D_t^{(\tilde{W})}\Pi(t)\\ &- g_x(t,x(t),y(t),z(t),\tilde{z}(t))\bar{p}(t), \end{aligned}$$

$$\Pi(t) = \frac{1}{2}\int_t^T (u(s) - M(s))^2 ds$$

and

$$\Phi(t,s) = \exp\left\{\int_t^s \left[A(r) - \frac{1}{\beta}\tilde{D}(r)\alpha_x(r,x(r))\right] dr\right\}.$$

According to Theorem 4.2 and (4.40), we have

Proposition 4.1. *Let $H_x(t)\Phi(t,s) \in \mathbb{L}_{1,2}(\mathbb{R})$, $0 \le t \le s \le T$. If $u(\cdot)$ is an optimal control strategy, then it is necessary to satisfy*

$$u(t) = M(t) - B(t)\mathbb{E}^u\left[\bar{q}(t)|\mathscr{F}_t^Y\right] - C(t)\mathbb{E}^u\left[D_t^{(W)}\bar{q}(t)|\mathscr{F}_t^Y\right], \tag{4.41}$$

where $\bar{q}(\cdot)$ is the solution of (4.40).

Note that a more explicit representation of (4.41) strongly depends on the specific structure of the distributions $\mathbb{E}^u\left[\bar{q}(t)|\mathscr{F}_t^Y\right]$ and $\mathbb{E}^u\left[D_t^{(W)}\bar{q}(t)|\mathscr{F}_t^Y\right]$. To illustrate this point, let us consider a special case of Proposition 4.1 in detail.

(H4.2) *Assume that g is independent of (x,y), and*

$$g(t,z,\tilde{z}) = c(t)z + \tilde{c}(t)\tilde{z}, f(x) = x \text{ and } \alpha(t,x) = \alpha(t),\ \forall (t,z,\tilde{z}) \in [0,T]\times\mathbb{R}^2,$$

where $c(\cdot)$, $\tilde{c}(\cdot)$, and $\alpha(\cdot)$ are deterministic and bounded.

It follows from (4.40) that

$$\bar{q}(t) = \bar{p}(T)\bar{A}(t), \quad D_t^{(W)}\bar{q}(t) = c(t)\bar{A}(t)\bar{p}(T),$$

with

$$\bar{A}(t) = -\int_t^T A(s)e^{\int_t^s A(r)dr}ds - 1.$$

Next, let

$$\hat{\bar{p}}_{s,t} = \mathbb{E}^u[\bar{p}(s)|\mathscr{F}_t^Y], \quad 0 \le t \le s \le T$$

be the optimal extrapolation of $\bar{p}(\cdot)$ with respect to

$$\mathscr{F}_t^Y = \sigma\{\tilde{W}(r); 0 \le r \le t\}.$$

Then (4.41) is rewritten as

$$u(t) = M(t) - \bar{A}(t)\Big(B(t) + c(t)C(t)\Big)\hat{\bar{p}}_{1,t}, \tag{4.42}$$

where

$$\hat{\bar{p}}_{s,t} = 1 + \int_0^t \bar{c}(r)\hat{\bar{p}}(r)d\tilde{W}(r)$$

with

$$\hat{\bar{p}}(r) = e^{\int_0^r \bar{c}(\theta)d\tilde{W}(\theta) - \frac{1}{2}\int_0^r \bar{c}^2(\theta)d\theta} \text{ and } \bar{c}(r) = \tilde{c}(r) - \frac{1}{\beta}\alpha(r) + \frac{1}{2}\beta.$$

Furthermore, the optimal cost functional can be derived in terms of (4.39) and (4.42).

We now summarize the result as follows.

Corollary 4.1. *Under (H4.2), the optimal control strategy of the underlying problem is given uniquely by (4.42).*

4.4 Notes

The earliest research on partially observable optimal control can be traced back to Florentin [22]. Since this paper was published in 1962, numerous people have contributed to this field. The interested reader is referred to Davis and Varaiya [16], Fleming and Pardoux [21], Bensoussan [6], Elliott et al. [20], Zhang and Xie [113], Tang [78], Shen et al. [72], and references cited therein for the development in various subjects, especially in maximum principle as well as dynamic programming.

However, prior to the beginning of 21st century, almost all the combined problems of control and filtering were formulated under the assumption that the state processes solve (forward) SDEs. With the rapid development and broad application of FBSDE in stochastic control theory, it is nature to say whether we can establish a combined model of filtering and control of FBSDE. Starting about from 2003, Zhen Wu and his graduate students at the School of Mathematics and System Sciences (now named the School of Mathematics), Shandong University, began to focus on exploring such a model. After about 5 years, the first result on Kalman–Bucy filtering of a special class of fully coupled FBSDEs was published, while a backward separation approach was proposed and was used to solve a partially observable LQ control problem driven by SDE in Wang and Wu [84]. At almost the same time, the first partially observable optimal control model of FBSDE was established by Wang and Wu [84] and Wu [100] from the viewpoint of mathematical finance, and then was studied by them via combining the backward separation approach with the maximum principle. Along this line, there are a few interesting papers to extend the model in several aspects, especially in maximum principle and nonlinear backward stochastic differential filtering equation. See, e.g., the doctoral dissertation of Wang [82], the survey paper of Wang et al. [93] for more details on these aspects. Note that how to obtain a dynamic programming principle corresponding to the partially observable forward-backward stochastic control model is also valuable topic. As far as we know, it has, however, not been explored so far.

The results introduced in this chapter are taken mainly from Wang et al. [88]. Similar to Chapter 3, some versions of verification theorem for optimality of Problem B can be derived. We omit them for the length of the book.

Chapter 5
LQ Optimal Control Models with Incomplete Information

In this chapter, we consider the so-called LQ problem with incomplete information aiming at obtaining more explicit results comparing with those of the previous chapters. We first consider this problem when the state is given by a linear FBSDE. After that we will specialize our results to the case when the state is governed by a BSDE only. In this case, explicit solution will be presented. Finally, we will apply our results to an optimal premium problem.

5.1 An LQ Model of FBSDE

As we introduced in the first chapter, Problem (FBLQ) we will study in this chapter consists of a linear FBSDE

$$\begin{cases} dx^v(t) = \big(a(t)x^v(t)+b(t)v(t)+\tilde{b}(t)\big)\,dt+c(t)dW(t)+\tilde{c}(t)d\tilde{W}(t), \\ -dy^v(t) = \big(A(t)x^v(t)+B(t)y^v(t)+C(t)z^v(t)+\tilde{C}(t)\tilde{z}^v(t) \\ \qquad\qquad +D(t)v(t)+\tilde{D}(t)\big)\,dt-z^v(t)dW(t)-\tilde{z}^v(t)d\tilde{W}(t), \\ x^v(0) = x_0, \quad y^v(T)=Fx^v(T)+G, \end{cases} \tag{5.1}$$

and the cost functional

$$\begin{aligned} J(v(\cdot)) = \frac{1}{2}\mathbb{E}\Big\{\int_0^T & \big[L(t)(x^v(t))^2+O(t)(y^v(t))^2+R(t)v^2(t) \\ & +2l(t)x^v(t)+2o(t)y^v(t)+2r(t)v(t)\big]\,dt \\ & +M(x^v(T))^2+2mx^v(T)+N(y^v(0))^2+2ny^v(0)\Big\}. \end{aligned} \tag{5.2}$$

G. Wang et al., *An Introduction to Optimal Control of FBSDE with Incomplete Information*, SpringerBriefs in Mathematics,
https://doi.org/10.1007/978-3-319-79039-8_5

The control process $v(\cdot)$ must be adapted to the information filtration derived from the observation process given by

$$\begin{cases} dY^v(t) = & (f(t)x^v(t)+g(t))dt+h(t)dW(t), \\ Y^v(0) = & 0. \end{cases} \tag{5.3}$$

5.1.1 Preliminary Results

Since the observation process depends on the control itself, we introduce the following decoupling technique. Define the processes $\big(x^0(\cdot),y^0(\cdot),z^0(\cdot),\tilde{z}^0(\cdot)\big)$, and $Y^0(\cdot)$ by

$$\begin{cases} dx^0(t) = & a(t)x^0(t)dt+c(t)dW(t)+\tilde{c}(t)d\tilde{W}(t), \\ -dy^0(t) = & \big(A(t)x^0(t)+B(t)y^0(t)+C(t)z^0(t)+\tilde{C}(t)\tilde{z}^0(t)\big)\,dt \\ & -z^0(t)dW(t)-\tilde{z}^0(t)d\tilde{W}(t), \\ x^0(0) = & x_0, \quad y^0(T)=Fx^0(T) \end{cases} \tag{5.4}$$

and

$$\begin{cases} dY^0(t) = & f(t)x^0(t)dt+h(t)dw(t), \\ Y^0(0) = & 0. \end{cases} \tag{5.5}$$

Let $v(\cdot)\in\mathscr{L}^2_{\mathbb{F}}(0,T;\mathbb{R})$ be a control process. Define $\big(x^1(\cdot),y^1(\cdot),z^1(\cdot),\tilde{z}^1(\cdot)\big)$, and $Y^1(\cdot)$ by

$$\begin{cases} \dot{x}^1(t) = & a(t)x^1(t)+b(t)v(t)+\tilde{b}(t), \\ -dy^1(t) = & \big(A(t)x^1(t)+B(t)y^1(t)+C(t)z^1(t)+\tilde{C}(t)\tilde{z}^1(t) \\ & +D(t)v(t)+\tilde{D}(t)\big)\,dt-z^1(t)dW(t)-\tilde{z}^1(t)d\tilde{W}(t), \\ x^1(0) = & 0, \quad y^1(T)=Fx^1(T)+G \end{cases} \tag{5.6}$$

and

$$\begin{cases} \dot{Y}^1(t) = & f(t)x^1(t)+g(t), \\ Y^1(0) = & 0. \end{cases} \tag{5.7}$$

Here the coefficients $a(\cdot)$, $b(\cdot)$, $\tilde{b}(\cdot)$, $c(\cdot)$, $\tilde{c}(\cdot)$, $f(\cdot)$, $g(\cdot)$, $h(\cdot)$, $h^{-1}(\cdot)$, $A(\cdot)$, $B(\cdot)$, $C(\cdot)$, $\tilde{C}(\cdot)$, $D(\cdot)$, and $\tilde{D}(\cdot)$ are uniformly bounded deterministic functions; x_0 and F are constants; and $\xi\in\mathscr{L}^2_{\mathscr{F}_T}(\Omega,\mathbb{R})$.

It is easy to see that (5.4), (5.5), (5.6), and (5.7) admit unique solutions, respectively. If we define

$$\begin{aligned} x^v(t) = & \; x^0(t)+x^1(t), \quad y^v(t)=y^0(t)+y^1(t), \quad z^v(t)=z^0(t)+z^1(t), \\ \tilde{z}^v(t) = & \; \tilde{z}^0(t)+\tilde{z}^1(t), \quad Y^v(t)=Y^0(t)+Y^1(t), \end{aligned} \tag{5.8}$$

it follows from Itô's formula and (5.4), (5.5), (5.6), (5.7), and (5.8) that $(x^v(\cdot)$, $y^v(\cdot),z^v(\cdot),\tilde{z}^v(\cdot))$ and $Y^v(\cdot)$ are the unique solutions of (5.1) and (5.3).

We will now derive a few results which will be useful when we derive the stochastic maximum principle for Problem (FBLQ). The first is about the identity of the filtration based on the observation with or without control. This is the first step in decoupling the filtering-control problem.

Lemma 5.1. *For any* $v(\cdot)\in\mathscr{U}_{ad}$, $\mathscr{F}_t^{Y^v}=\mathscr{F}_t^{Y^0}$.

Proof. For any $v(\cdot)\in\mathscr{U}_{ad}$, since $v(t)$ is $\mathscr{F}_t^{Y^0}$-adapted, then it follows from (5.4) that $x^1(t)$ is $\mathscr{F}_t^{Y^0}$-adapted, so is $Y^1(t)$. Then $Y^v(t)=Y^0(t)+Y^1(t)$ is $\mathscr{F}_t^{Y^0}$-adapted, i.e., $\mathscr{F}_t^{Y^v}\subseteq\mathscr{F}_t^{Y^0}$. In a similar way, we get $\mathscr{F}_t^{Y^0}\subseteq\mathscr{F}_t^{Y^v}$ via the equality $Y^0(t)=Y^v(t)-Y^1(t)$. The proof is thus complete. □

The following estimates describe the continuity of state with respect to control, which are derived by Itô's formula and Gronwall's inequality. See also [19, 109] for similar arguments.

Lemma 5.2. *For any* $v_i(\cdot)\in\mathscr{L}^2_{\mathbb{F}^{W,\tilde{W}}}(0,T;\mathbb{R})$, *let* $(x^{v_i}(\cdot),y^{v_i}(\cdot),z^{v_i}(\cdot),\tilde{z}^{v_i}(\cdot))$ *be the solution of (1.28) corresponding to* $v_i(\cdot)$ $(i=1,2)$. *Then there is a constant* $C_0>0$ *such that*

$$\begin{aligned}\sup_{0\le t\le T}\mathbb{E}|x^{v_1}(t)-x^{v_2}(t)|^2\le&C_0\mathbb{E}\int_0^T|v_1(t)-v_2(t)|^2dt,\\ \sup_{0\le t\le T}\mathbb{E}|y^{v_1}(t)-y^{v_2}(t)|^2\le&C_0\Big[\mathbb{E}|x^{v_1}(T)-x^{v_2}(T)|^2\\ &+\int_0^T\sup_{0\le s\le t}\mathbb{E}|x^{v_1}(s)-x^{v_2}(t)|^2dt\\ &+\mathbb{E}\int_0^T|v_1(t)-v_2(t)|^2dt\Big].\end{aligned}$$

The next result is the key in decoupling the filtering-control problem.

Lemma 5.3.

$$\inf_{v'(\cdot)\in\mathscr{U}_{ad}}J(v'(\cdot))=\inf_{v(\cdot)\in\mathscr{U}^0_{ad}}J(v(\cdot)).$$

Proof. From Definition 1.3, we have $\mathscr{U}_{ad}\subseteq\mathscr{U}^0_{ad}$, and thus,

$$\inf_{v'(\cdot)\in\mathscr{U}_{ad}}J(v'(\cdot))\ge\inf_{v(\cdot)\in\mathscr{U}^0_{ad}}J(v(\cdot)).$$

In what follows, we prove that the reverse inequality holds by three steps.

Step 1: $\mathscr{U}_{ad}$ is dense in $\mathscr{U}^0_{ad}$ under the metric of $\mathscr{L}^2_{\mathbb{F}^{Y^0}}(0,T;\mathbb{R})$.

For any $v(\cdot)\in\mathscr{U}^0_{ad}$, define

$$v_k(t)=\begin{cases}v(0), & \text{for } 0\le t\le\delta_k,\\ \frac{1}{\delta_k}\int_{(i-1)\delta_k}^{i\delta_k}v(s)ds, & \text{for } i\delta_k<t\le(i+1)\delta_k,\end{cases}$$

where $v(0) \in \mathbb{R}$, i, k are natural numbers, $1 \le i \le k-1$, and $\delta_k = T/k$. Then $v_k(t)$ is $\mathscr{F}^{Y^0}_{i\delta_k}$-adapted for any $i\delta_k < t \le (i+1)\delta_k$, and for any k,

$$\sup_{0\le t\le T} |v_k(t)| \le |v(0)| + \sup_{0\le t\le T} |v(t)|.$$

Thus, $v_k(\cdot) \in \mathscr{U}^0_{ad}$. Let $(x^{v_k}(\cdot), y^{v_k}(\cdot), z^{v_k}(\cdot), \tilde{z}^{v_k}(\cdot))$ and $Y^{v_k}(\cdot)$ be the trajectories of (5.1) and (5.3) corresponding to $v_k(\cdot)$. Similar to [9], from (5.6), (5.7), and the last equality of (5.8), we verify that $v_k(\cdot)$ is adapted to $\mathscr{F}^{Y^0}_t$ and $\mathscr{F}^{Y^{v_k}}_t$, and $\mathscr{F}^{Y^0}_t = \mathscr{F}^{Y^{v_k}}_t$. Then $v_k(\cdot)$ belongs to $\mathscr{U}_{ad}$, and thus, (5.1) has a unique solution $(x^{v_k}(\cdot), y^{v_k}(\cdot), z^{v_k}(\cdot), \tilde{z}^{v_k}(\cdot)) \in \mathscr{L}^2_{\mathbb{F}^{W,\tilde{W}}}(0,T;\mathbb{R}^4)$. On the other hand, $v_k(\cdot) \to v(\cdot)$ in probability when $k \to +\infty$. Using again the integrability condition of $v(\cdot)$ in $\mathscr{U}^0_{ad}$, we derive $v_k(\cdot) \to v(\cdot)$ as $k \to +\infty$ in $\mathscr{L}^2_{\mathbb{F}^{Y^0}}(0,T;\mathbb{R})$, i.e., $\mathscr{U}_{ad}$ is dense in $\mathscr{U}^0_{ad}$.

Step 2:

$$\lim_{k\to+\infty} J(v_k(\cdot)) = J(v(\cdot)),$$

where $v(\cdot)$, $v_k(\cdot)$ and $(x^{v_k}(\cdot), y^{v_k}(\cdot), z^{v_k}(\cdot), \tilde{z}^{v_k}(\cdot))$ are defined as in Step 1.

By (5.2) and Hölder's inequality, it yields

$$\begin{aligned}
&2|J(v_k(\cdot)) - J(v(\cdot))| \\
&\le \sqrt{\mathbb{E}\int_0^T |L(t)(x^{v_k}(t) + x^v(t)) + 2l(t)|^2 dt}\sqrt{\mathbb{E}\int_0^T |x^{v_k}(t) - x^v(t)|^2 dt} \\
&+ \sqrt{\mathbb{E}\int_0^T |O(t)(y^{v_k}(t) + y^v(t)) + 2o(t)|^2 dt}\sqrt{\mathbb{E}\int_0^T |y^{v_k}(t) - y^v(t)|^2 dt} \\
&+ \sqrt{\mathbb{E}\int_0^T |R(t)(v_k(t) + v(t)) + 2r(t)|^2 dt}\sqrt{\mathbb{E}\int_0^T |v_k(t) - v(t)|^2 dt} \\
&+ \sqrt{\mathbb{E}|M(x^{v_k}(T) + x^v(T)) + 2m|^2}\sqrt{\mathbb{E}|x^{v_k}(T) - x^v(T)|^2} \\
&+ \sqrt{\mathbb{E}|N(y^{v_k}(0) + y^v(0)) + 2n|^2}\sqrt{\mathbb{E}|y^{v_k}(0) - y^v(0)|^2}.
\end{aligned}$$

Then Lemma 5.2 implies that $J(v_k(\cdot)) \to J(v(\cdot))$ when k goes to $+\infty$.

Step 3:

$$\inf_{v'(\cdot)\in\mathscr{U}_{ad}} J(v'(\cdot)) \le \inf_{v(\cdot)\in\mathscr{U}^0_{ad}} J(v(\cdot)).$$

Since $v_k(\cdot) \in \mathscr{U}_{ad}$, then

$$\inf_{v'(\cdot)\in\mathscr{U}_{ad}} J(v'(\cdot)) \le J(v_k(\cdot)),$$

and consequently, $\inf_{v'(\cdot)\in\mathscr{U}_{ad}} J(v'(\cdot)) \le J(v(\cdot))$ by sending $k \to +\infty$. Due to the arbitrariness of $v(\cdot)$, the desired inequality holds. Thus, the proof is complete. □

5.1.2 Optimality Condition

We first establish a necessary condition and then a sufficient condition for optimality of Problem (FBLQ). According to Lemma 5.3, it suffices to study the optimality of $J[v]$ over $\mathscr{U}_{ad}^0$. In addition, $\mathscr{U}_{ad}^0$ is fixed (i.e., independent of control or state), and it is more convenient to get the optimality conditions in $\mathscr{U}_{ad}^0$ by variational method. Note that these results are different from the existing literature, say, [31, 54, 61, 75, 96, 88, 103], mainly due to the fact that the drift coefficient of the observation equation is linear with respect to the state, and the state noise is correlated to the observation noise.

Theorem 5.1. *Suppose that $u(\cdot)$ is an optimal control of Problem (FBLQ), in the sense that*

$$\frac{d}{d\varepsilon}J(u(\cdot)+\varepsilon v(\cdot))|_{\varepsilon=0}=0 \textit{ for any } v(\cdot)+u(\cdot)\in\mathscr{U}_{ad},$$

and $(x(\cdot),y(\cdot),z(\cdot),\tilde{z}(\cdot))$ is the corresponding optimal state. Then FBSDE

$$\begin{cases} dp(t)=(B(t)p(t)-O(t)y(t)-o(t))dt+C(t)p(t)dW(t)+\tilde{C}(t)p(t)d\tilde{W}(t),\\ -dq(t)=(a(t)q(t)-A(t)p(t)+L(t)x(t)+l(t))dt-k(t)dW(t)-\tilde{k}(t)d\tilde{W}(t),\\ p(0)=-Ny(0)-n,\quad q(T)=-Fp(T)+Mx(T)+m \end{cases} \tag{5.9}$$

admits a unique solution $(p(\cdot),q(\cdot),k(\cdot),\tilde{k}(\cdot))\in\mathscr{L}^2_{\mathbb{F}^{W,\tilde{W}}}(0,T;\mathbb{R}^4)$ such that

$$R(t)u(t)-D(t)\mathbb{E}\left[p(t)\middle|\mathscr{F}_t^Y\right]+b(t)\mathbb{E}\left[q(t)\middle|\mathscr{F}_t^Y\right]+r(t)=0 \tag{5.10}$$

with

$$\mathscr{F}_t^Y=\sigma\{Y^u(s);0\le s\le t\}.$$

Proof. According to Lemma 5.3, if $u(\cdot)$ is an optimal control, then

$$J(u(\cdot))=\inf_{v(\cdot)\in\mathscr{U}_{ad}^0}J(v(\cdot)).$$

For $v(\cdot)\in\mathscr{U}_{ad}$, we introduce a variational equation

$$\begin{cases} \dot{x}_1(t)=a_t x_1(t)+b(t)v(t),\\ -dy_1(t)=(A(t)x_1(t)+B(t)y_1(t)+C(t)z_1(t)+\tilde{C}(t)\tilde{z}_1(t)+D(t)v(t))dt\\ \qquad -z_1(t)dW(t)-\tilde{z}_1(t)d\tilde{W}(t),\\ x_1(0)=0,\quad y_1(T)=Fx_1(T), \end{cases}$$

which admits a unique solution $(x_1(\cdot),y_1(\cdot),z_1(\cdot),\tilde{z}_1(\cdot))\in\mathscr{L}^2_{\mathbb{F}^{W,\tilde{W}}}(0,T;\mathbb{R}^4)$. By the optimality of $u(\cdot)$ using the first variation of $J(v(\cdot))$ with Lemma 4.4, we have

$$
\begin{aligned}
0 =& \frac{d}{d\varepsilon} J(u(\cdot)+\varepsilon v(\cdot))|_{\varepsilon=0} \\
=& \mathbb{E}\left\{ \int_0^T [(L(t)x(t)+l(t))x_1(t)+(O(t)y(t)+o(t))y_1(t) \right. \\
&+ (R(t)u(t)+r(t))v(t)]dt \\
&\left. + (Mx(T)+m)x_1(T)+(Ny(0)+n)y_1(0) \right\}.
\end{aligned} \tag{5.11}
$$

On the other hand, once $(x(\cdot),y(\cdot),z(\cdot),\tilde{z}(\cdot))$ is determined, (5.9) admits a unique solution $\left(p(\cdot),q(\cdot),k(\cdot),\tilde{k}(\cdot)\right) \in \mathscr{L}^2_{\mathbb{F}^{W,\tilde{W}}}\left(0,T;\mathbb{R}^4\right)$. It follows from Itô's formula in a form of differentiation by parts of the product of two stochastic processes that

$$
\begin{aligned}
d(p(t)y_1(t)) =& -[(O(t)y(t)+o(t))y_1(t)+(A(t)x_1(t)+D(t)v(t))p(t)]\,dt \\
&+ p(t)(C(t)y_1(t)+z_1(t))dW(t)+p(t)(\tilde{C}(t)y_1(t)+\tilde{z}_1(t))d\tilde{W}(t).
\end{aligned}
$$

Thus,

$$
\begin{aligned}
&\mathbb{E}\left[Fp(T)x_1(T)+(Ny(0)+n)y_1(0)\right] \\
=& -\mathbb{E}\int_0^T [(O(t)y(t)+o(t))y_1(t)+(A(t)x_1(t)+D(t)v(t))p(t)]\,dt.
\end{aligned} \tag{5.12}
$$

Similarly, by differentiation by parts again,

$$
\begin{aligned}
d(q(t)x_1(t)) =& \{[A(t)p(t)-(L(t)x(t)+l(t))]\,x_1(t)+b(t)q(t)v(t)\}\,dt \\
&+ k(t)x_1(t)dW(t)+\tilde{k}(t)x_1(t)d\tilde{W}(t),
\end{aligned}
$$

and then

$$
\begin{aligned}
&\mathbb{E}\left[(Mx(T)+m-Fp(T))x_1(T)\right] \\
=& \mathbb{E}\int_0^T \{[A(t)p(t)-(L(t)x(t)+l(t))]\,x_1(t)+b(t)q(t)v(t)\}\,dt.
\end{aligned} \tag{5.13}
$$

Recall that $R(t)$, $D(t)$, $b(t)$, and $r(t)$ are deterministic, and $u(t)$ and $v(t)$ are $\mathscr{F}_t^{Y^0}$-adapted. Substituting (5.12) and (5.13) into (5.11), we get

$$
\begin{aligned}
0 =& \mathbb{E}\int_0^T (R(t)u(t)-D(t)p(t)+b(t)q(t)+r(t))v(t)dt \\
=& \mathbb{E}\int_0^T \Big(R(t)u(t)-D(t)\mathbb{E}\left[p(t)\middle|\mathscr{F}_t^{Y^0}\right] \\
&+ b(t)\mathbb{E}\left[q(t)\middle|\mathscr{F}_t^{Y^0}\right]+r(t)\Big)\, v(t)dt.
\end{aligned}
$$

Hence,

$$
R(t)u(t)-D(t)\mathbb{E}\left[p(t)\middle|\mathscr{F}_t^{Y^0}\right]+b(t)\mathbb{E}\left[q(t)\middle|\mathscr{F}_t^{Y^0}\right]+r(t)=0.
$$

Furthermore, since $u(\cdot) \in \mathscr{U}_{ad}$, it follows from Lemma 5.1 that $\mathscr{F}_t^{Y^0} = \mathscr{F}_t^Y$, and thus the desired conclusion. □

We now study the sufficiency of the above result. Introduce an FBSDE with (5.10)

$$\begin{cases} dx(t) = (a(t)x(t) + b(t)u(t) + \tilde{b}(t))dt + c(t)dW(t) + \tilde{c}(t)d\tilde{W}(t), \\ -dy(t) = (A(t)x(t) + B(t)y(t) + C(t)z(t) + \tilde{C}(t)\tilde{z}(t) + D(t)u(t) + \tilde{D}(t))dt \\ \qquad - z(t)dW(t) - \tilde{z}(t)d\tilde{W}(t), \\ dp(t) = (B(t)p(t) - O(t)y(t) - o(t))dt + C(t)p(t)dW(t) + \tilde{C}(t)p(t)d\tilde{W}(t), \\ -dq(t) = (a(t)q(t) - A(t)p(t) + L(t)x(t) + l(t))dt - k(t)dW(t) - \tilde{k}(t)d\tilde{W}(t), \\ x(0) = x_0, \quad y(T) = Fx(T) + G, \\ p(0) = -Ny(0) - n, \quad q(T) = -Fp(T) + Mx(T) + m, \end{cases} \tag{5.14}$$

which is called a generalized stochastic Hamiltonian system in the field of Pontryagin's maximum principle. If a process $(x(\cdot), y(\cdot), z(\cdot), \tilde{z}(\cdot), p(\cdot), q(\cdot), k(\cdot), \tilde{k}(\cdot)) \in \mathscr{L}^2_{\mathbb{F}^{W,\tilde{W}}}(0,T;\mathbb{R}^8)$ satisfies (5.14), we call it an (adapted) solution of (5.14).

Theorem 5.2. *Let $u(\cdot) \in \mathscr{U}_{ad}$ satisfy*

$$R(t)u(t) - D(t)\mathbb{E}\left[p(t)\middle|\mathscr{F}_t^Y\right] + b(t)\mathbb{E}\left[q(t)\middle|\mathscr{F}_t^Y\right] + r(t) = 0,$$

where $(x(\cdot), y(\cdot), z(\cdot), \tilde{z}(\cdot), p(\cdot), q(\cdot), k(\cdot), \tilde{k}(\cdot))$ is a solution to (5.14). Then u is an optimal control of Problem (FBLQ).

Proof. For any admissible control $v(\cdot)$, the total variation is

$$J(v(\cdot)) - J(u(\cdot)) = I + II \tag{5.15}$$

with $(x(\cdot), y(\cdot)) = (x^u(\cdot), y^u(\cdot))$, the pure quadratic part is

$$\begin{aligned} I = \frac{1}{2}\mathbb{E}\Big\{ \int_0^T & \left[L(t)(x^v(t) - x(t))^2 + O(t)(y^v(t) - y(t))^2 \right. \\ & \left. + R(t)(v(t) - u(t))^2\right] dt + M(x^v(t) - x(t))^2 + N(y^v(0) - y(0))^2 \Big\}, \end{aligned}$$

and the quasi-linear part is

$$\begin{aligned} II = \mathbb{E}\Big\{ \int_0^T & [(L(t)x(t) + l(t))(x^v(t) - x(t)) \\ & + (O(t)y(t) + o(t))(y^v(t) - y(t)) + (R(t)u(t) + r(t))(v(t) - u(t))]dt \\ & + (Mx(t) + m)(x^v(t) - x(t)) + (Ny + n)(y^v(0) - y(0)) \Big\}. \end{aligned} \tag{5.16}$$

Note that $I \geq 0$ holds for any admissible control $v(\cdot)$. Then it is enough to prove that

$$II = 0.$$

It follows from Itô's formula that

$$\begin{aligned} d\left[p(t)(y^v(t)-y(t))\right] = &-\{(O(t)y(t)+o(t))(y^v(t)-y(t)) \\ &+p(t)\left[A(t)(x^v(t)-x(t))+D(t)(v(t)-u(t))\right]\}dt \\ &+p(t)\left[C(t)(y^v(t)-y(t))+z^v(t)-z(t)\right]dW(t) \\ &+p(t)\left[\tilde{C}(t)(y^v(t)-y(t))+\tilde{z}^v(t)-\tilde{z}(t)\right]d\tilde{W}(t). \end{aligned}$$

Thus, by using differentiation by parts, similar to (5.12), we get

$$\begin{aligned} &\mathbb{E}\left[Fp(t)(x^v(t)-x(t))+(n+Ny_0)(y^v(0)-y(0))\right] \\ &= -\mathbb{E}\int_0^T \{(O(t)y(t)+o(t))(y^v(t)-y(t)) \\ &+p(t)\left[A(t)(x^v(t)-x(t))+D(t)(v(t)-u(t))\right]\}dt. \end{aligned} \tag{5.17}$$

Similarly by differentiation by parts, similar to (5.13), we obtain,

$$\begin{aligned} d\left[q(t)(x^v(t)-x(t))\right] = &\left[(A(t)p(t)-L(t)x(t)-l(t))(x^v(t)-x(t))\right. \\ &\left.+b(t)q(t)(v(t)-u(t))\right]dt \\ &+k(t)(x^v(t)-x(t))dW(t)+\tilde{k}(t)(x^v(t)-x(t))d\tilde{W}(t), \end{aligned}$$

and then

$$\begin{aligned} &\mathbb{E}\left[(Mx(t)-Fp(t)+m)(x^v(t)-x(t))\right] \\ &= \mathbb{E}\int_0^T \left[b(t)q(t)(v(t)-u(t))\right. \\ &\left.+(A(t)p(t)-L(t)x(t)-l(t))(x^v(t)-x(t))\right]dt. \end{aligned} \tag{5.18}$$

Plugging (5.17) and (5.18) into (5.16) and using Lemma 5.1, we have

$$\begin{aligned} II &= \mathbb{E}\int_0^T (R(t)u(t)-D(t)p(t)+b(t)q(t)+r(t))(v(t)-u(t))dt \\ &= \mathbb{E}\int_0^T \left(R(t)u(t)-D(t)\mathbb{E}\left[p(t)\middle|\mathscr{F}_t^{Y^0}\right]+b(t)\mathbb{E}\left[q(t)\middle|\mathscr{F}_t^{Y^0}\right]+r(t)\right) \\ &\quad \times (v(t)-u(t))dt \\ &= 0. \end{aligned}$$

Then the proof is complete. □

Corollary 5.1. *Let $R(\cdot)$ and $R^{-1}(\cdot)$ be uniformly bounded, deterministic functions. If $u(\cdot)$ is an optimal control of Problem (FBLQ), then $u(\cdot)$ is unique.*

Proof. Let $u(\cdot)$ and $\bar{u}(\cdot)$ be both optimal controls of Problem (FBLQ) with the same optimum, and let $(x(\cdot), y(\cdot), z(\cdot), \tilde{z}(\cdot))$ and $(\bar{x}(\cdot), \bar{y}(\cdot), \bar{z}(\cdot), \bar{\tilde{z}}(\cdot))$ be the corresponding optimal states. It is easy to see that $((x(\cdot)+\bar{x}(\cdot))/2, (y(\cdot)+\bar{y}(\cdot))/2, (z(\cdot)+\bar{z}(\cdot))/2, (\tilde{z}(\cdot)+\bar{\tilde{z}}(\cdot))/2)$ is the state corresponding to $(u(\cdot)+\bar{u}(\cdot))/2$. Then

$$
\begin{aligned}
2J(u(\cdot)) &= J(u(\cdot)) + J(\bar{u}(\cdot)) \\
&= 2J\left(\frac{u(\cdot)+\bar{u}(\cdot)}{2}\right) + \mathbb{E}\left\{\int_0^T \left[L(t)\left(\frac{x(t)-x^{\bar{u}}(t)}{2}\right)^2\right.\right. \\
&\quad \left. + O(t)\left(\frac{y(t)-y^{\bar{u}}(t)}{2}\right)^2 + R(t)\left(\frac{u(t)-\bar{u}(t)}{2}\right)^2\right] dt \\
&\quad \left. + M\left(\frac{x(t)-x^{\bar{u}}(t)}{2}\right)^2 + N\left(\frac{y(0)-y^{\bar{u}}(0)}{2}\right)^2\right\} \\
&\geq 2J(u(\cdot)) + \mathbb{E}\int_0^T R(t)\left(\frac{u(t)-\bar{u}(t)}{2}\right)^2 dt.
\end{aligned}
$$

Since $R(t) > 0$, we have $u(\cdot) = \bar{u}(\cdot)$. □

5.1.3 Filtering

It follows from (5.10) that

$$
u(t) = \frac{1}{R(t)}\left(D(t)\mathbb{E}\left[p(t)\middle|\mathscr{F}_t^Y\right] - b(t)\mathbb{E}\left[q(t)\middle|\mathscr{F}_t^Y\right] - r(t)\right).
$$

This shows that it is necessary to compute the optimal filtering of $(p(t), q(t))$ based on $\mathscr{F}_t^Y$. Furthermore, since $(p(\cdot), q(\cdot))$ is closely related to $(x(\cdot), y(\cdot))$, we need to analyze the optimal filtering of FBSDEs (5.1) and (5.9). The earliest work on filtering for FBSDEs was traced back to Wang and Wu [84], where a Feynman–Kac formula method was used to calculate the filtering of linear FBSDEs. Note that it seems that the Feynman–Kac formula method does not work here, mainly due to the existence of the control $v(\cdot)$.

Recall that $Y^v(\cdot)$ governed by (5.3) is the observation. For any $v(\cdot) \in \mathscr{U}_{ad}$, let

$$
\hat{\xi}(t) = \mathbb{E}\left[\xi(t)\middle|\mathscr{F}_t^{Y^v}\right], \tag{5.19}
$$

with $\xi(t) = x^0(t), x^v(t), y^v(t), z^v(t), \tilde{z}^v(t), p(t), q(t), k(t), x^v(t)y^v(t)$,

$$
\hat{G} = \mathbb{E}\left[G\middle|\mathscr{F}_t^{Y^v}\right] \text{ and } P(t) = \mathbb{E}(x^v(t) - \hat{x}^v(t))^2
$$

be the optimal filtering and the mean square error of $\xi(t)$, G and $\hat{x}^v(t)$, respectively.

We state a filtering result of (5.1), which plays an important role in representing optimal control.

Lemma 5.4. *For any $v(\cdot) \in \mathscr{U}_{ad}$, the optimal filtering $(\hat{x}^v(t), \hat{y}^v(t), \hat{z}^v(t), \hat{\tilde{z}}^v(t))$ of the solution $(x^v(t), y^v(t), z^v(t), \tilde{z}^v(t))$ to (5.1) with respect to $\mathscr{F}_t^{Y^v}$ satisfies*

$$\begin{cases} d\hat{x}^v(t) = \left(a(t)\hat{x}^v(t) + b(t)v(t) + \tilde{b}(t)\right)dt + \left(c(t) + \dfrac{P(t)f(t)}{h(t)}\right)d\hat{W}(t), \\ -d\hat{y}^v(t) = \left(A(t)\hat{x}^v(t) + B(t)\hat{y}^v(t) + C(t)\hat{z}^v(t) + \tilde{C}(t)\hat{\tilde{z}}^v(t) + D(t)v(t) + \tilde{D}(t)\right)dt \\ \qquad\qquad - \hat{Z}^v(t)d\hat{W}(t), \\ \hat{x}_0^v = x_0, \quad \hat{y}^v(T) = F\hat{x}^v(T) + \tilde{G}, \end{cases} \tag{5.20}$$

where the mean square error $P(t)$ of the estimate $\hat{x}^v(t)$ is the unique solution of

$$\begin{cases} \dot{P}(t) - 2a(t)P(t) + \left(c(t) + \dfrac{P(t)f(t)}{h(t)}\right)^2 - (c(t) + \tilde{c}(t))^2 = 0, \\ P_0 = 0, \end{cases} \tag{5.21}$$

$$\begin{aligned} \hat{W}(t) &= \int_0^t \frac{1}{h(s)}[dY(s) - (f(s)\hat{x}^v(s) + g(s))ds] \\ &= \int_0^t \frac{f(s)}{h(s)}(x^v(s) - \hat{x}^v(s))ds + W(t) \end{aligned} \tag{5.22}$$

is a standard Brownian motion with values in $\mathbb{R}$, and

$$\hat{Z}^v(t) = \hat{z}^v(t) + \frac{f(t)}{h(t)}\left(\widehat{x^v(t)y^v(t)} - \hat{x}^v(t)\hat{y}^v(t)\right). \tag{5.23}$$

We highlight that $\hat{Z}^v(t)$ defined by (5.23) is a part of solution $(\hat{y}^v(t), \hat{Z}^v(t))$ to the BSDE in (5.20), which can be computed by the Malliavin derivative of $\hat{y}^v(t)$ with respect to $\hat{W}(t)$ under some standard conditions [19].

5.1.4 Feedback

In this section, we do our best to give a feedback form of the optimal control $u(\cdot)$. We assume that $O(t) = 0$ for simplicity. It implies that the running cost part of (5.2) does not include the quadratic term of the state $y(\cdot)$. Similar to (5.19), let

$$\begin{aligned} &\widehat{x(t)q(t)} = \mathbb{E}\left[x(t)q(t)\middle|\mathscr{F}_t^Y\right], \quad \widehat{x^{\mathbf{m}}(t)} = \mathbb{E}\left[x^{\mathbf{m}}(t)\middle|\mathscr{F}_t^Y\right], \\ &\widehat{x^{\mathbf{m}}(t)p(t)} = \mathbb{E}\left[x^{\mathbf{m}}(t)p(t)\middle|\mathscr{F}_t^Y\right], \quad \mathbf{m} = 1, 2, 3, \cdots \end{aligned}$$

be the optimal filters of $x(t)q(t)$, $x^{\mathbf{m}}(t)$, $x^{\mathbf{m}}(t)p(t)$ based on the observation $Y(\cdot)$ up to time t, respectively.

We now state a filtering result of the adjoint equation of Problem (FBLQ).

Lemma 5.5. *Let $O(t) = 0$ hold. The optimal filtering of $(p(t), q(t), k(t))$ depending on $\mathscr{F}_t^Y$ satisfies*

$$\begin{cases} d\hat{p}(t) = (B(t)\hat{p}(t) - o(t))dt + \left[C(t)\hat{p}(t) + \dfrac{f(t)}{h(t)}\left(\widehat{x(t)p(t)} - \hat{x}(t)\hat{p}(t)\right)\right] d\hat{W}(t), \\ -d\hat{q}(t) = (a(t)\hat{q}(t) - A(t)\hat{p}(t) + L(t)\hat{x}(t) + l(t))dt - \hat{K}(t)d\hat{W}(t), \\ \hat{p}(0) = -Ny(0) - n, \quad \hat{q}(T) = M\hat{x}(T) - F\hat{p}(T) + m \end{cases} \tag{5.24}$$

with

$$\hat{K}(t) = \hat{k}(t) + \frac{f(t)}{h(t)}\left(\widehat{x(t)q(t)} - \hat{x}(t)\hat{q}(t)\right),$$

where $(\hat{x}(\cdot), \hat{y}(\cdot))$, $\hat{W}(\cdot)$, and $\widehat{x^{\mathbf{m}}(\cdot)p(\cdot)}$ satisfy (5.20) with $v(\cdot) = u(\cdot)$, (5.22), and

$$\begin{cases} d\widehat{x^{\mathbf{m}}(t)p(t)} = \Big[(\mathbf{m}a(t) + B(t))\widehat{x^{\mathbf{m}}(t)p(t)} - o(t)\widehat{x^{\mathbf{m}}(t)} \\ \qquad + \mathbf{m}\left(b(t)u(t) + \bar{b}(t) + c(t)C(t) + \bar{c}(t)\bar{C}(t)\right)\widehat{x^{\mathbf{m}-1}(t)p(t)}\Big]dt \\ \qquad + \Big[\mathbf{m}c(t)\widehat{x^{\mathbf{m}-1}(t)p(t)} + C(t)\widehat{x^{\mathbf{m}}(t)p(t)} \\ \qquad + \dfrac{f(t)}{h(t)}\left(\widehat{x^{\mathbf{m}+1}(t)p(t)} - \hat{x}(t)\widehat{x^{\mathbf{m}}(t)p(t)}\right)\Big]d\hat{W}(t), \\ \widehat{x^{\mathbf{m}}(0)p(0)} = -x_0^{\mathbf{m}}(Ny(0) + n), \quad \mathbf{m} = 1, 2, 3, \cdots, \end{cases}$$

respectively.

Theorem 5.3. *Let $O(t) = 0$ hold. If*

$$u(t) = \frac{1}{R(t)}(D(t)\hat{p}(t) - b(t)\hat{q}(t) - r(t))$$

is the optimal control of Problem (FBLQ), then it can be represented as

$$u(t) = \frac{1}{R(t)}[(D(t) - b(t)\Sigma(t))\hat{p}(t) - b(t)\pi(t)\hat{x}(t) - b(t)\theta(t) - r(t)],$$

where $(\hat{x}(\cdot), \hat{y}(\cdot), \hat{z}(\cdot), \hat{\tilde{z}}(\cdot))$, $(\hat{p}(\cdot), \hat{q}(\cdot), \hat{k}(\cdot))$, $\pi(\cdot)$, $\Sigma(\cdot)$, and $\theta(\cdot)$ are the solutions of (5.20) with $v(\cdot) = u(\cdot)$, (5.24), (5.28), (5.29), and (5.30), respectively.

Proof. For any $v(\cdot) \in \mathscr{U}_{ad}$, (5.1) admits a unique solution $(x^v(\cdot), y^v(\cdot), z^v(\cdot), \tilde{z}^v(\cdot))$, and consequently, (5.9) admits a unique solution $(p(\cdot), q(\cdot), k(\cdot), \tilde{k}(\cdot))$. Noting the terminal condition of (5.9), we set

$$q(t) = \pi(t)x(t) + \Sigma(t)p(t) + \theta(t) \text{ with } \pi(T) = M, \Sigma(T) = -F \text{ and } \theta(T) = m, \tag{5.25}$$

where $\pi(\cdot)$, $\Sigma(\cdot)$, and $\theta(\cdot)$ are deterministic and differentiable functions. Substituting (5.25) into

$$u(t)=\frac{1}{R(t)}(D(t)\hat{p}(t)-b(t)\hat{q}(t)-r(t))$$

and applying Itô's formula to (5.25), we obtain

$$\begin{aligned}dq(t)&=\dot{\pi}(t)x(t)dt+\pi(t)dx(t)+\dot{\Sigma}(t)p(t)dt+\Sigma(t)dp(t)+\dot{\theta}(t)dt\\&=\Big\{\dot{\pi}(t)x(t)+\pi(t)\Big[a(t)x(t)+\frac{1}{R(t)}b(t)((D(t)-b(t)\Sigma(t))\hat{p}(t)\\&\quad-b(t)\pi(t)\hat{x}(t)-b(t)\theta(t)-r(t))+\tilde{b}(t)\Big]\\&\quad+\dot{\Sigma}(t)p(t)+\Sigma(t)(B(t)p(t)-o(t))+\dot{\theta}(t)\Big\}dt\\&\quad+(c(t)\pi(t)+C(t)p(t)\Sigma(t))dW(t)+(\tilde{c}(t)\pi(t)+\tilde{C}(t)p(t)\Sigma(t))d\tilde{W}(t).\end{aligned}$$

Comparing the above equality with (5.9), it yields

$$k(t)=c(t)\pi(t)+C(t)p(t)\Sigma(t),\quad \tilde{k}(t)=\tilde{c}(t)\pi(t)+\tilde{C}(t)p(t)\Sigma(t)$$

and

$$\begin{aligned}&\dot{\pi}(t)x(t)+\pi(t)\Big\{a(t)x(t)+\frac{1}{R(t)}b(t)[(D(t)-b(t)\Sigma(t))\hat{p}(t)-b(t)\pi(t)\hat{x}(t)\\&-b(t)\theta(t)-r(t)]+\tilde{b}(t)\Big\}+\dot{\Sigma}(t)p(t)+\Sigma(t)(B(t)p(t)-o(t))+\dot{\theta}(t)\\=&-(a(t)q(t)-A(t)p(t)+L(t)x(t)+l(t))\\=&-(a(t)\pi(t)+L(t))x(t)-(a(t)\Sigma(t)-A(t))p(t)-a(t)\theta(t)-l(t).\end{aligned}\tag{5.26}$$

Taking the conditional expectation $\mathbb{E}\left[\cdot\middle|\mathscr{F}^Y(t)\right]$ on both sides of (5.26),

$$\begin{aligned}&\dot{\pi}(t)\hat{x}(t)+\pi(t)\Big[a(t)\hat{x}(t)+\frac{1}{R(t)}b(t)((D(t)-b(t)\Sigma(t))\hat{p}(t)-b(t)\pi(t)\hat{x}(t)\\&-b(t)\theta(t)-r(t))+\tilde{b}(t)\Big]+\dot{\Sigma}(t)\hat{p}(t)+\Sigma(t)(B(t)\hat{p}(t)-o(t))+\dot{\theta}(t)\\=&-(a(t)\pi(t)+L(t))\hat{x}(t)-(a(t)\Sigma(t)-A(t))\hat{p}(t)-a(t)\theta(t)-l(t).\end{aligned}\tag{5.27}$$

Comparing the coefficients of $\hat{x}(t)$ and $\hat{p}(t)$ in (5.27), we derive the following Riccati equations:

$$\begin{cases}\dot{\pi}(t)+2a(t)\pi(t)-\dfrac{1}{R(t)}b^2(t)\pi^2(t)+L(t)=0,\\ \pi(T)=M,\end{cases}\tag{5.28}$$

$$\begin{cases}\dot{\Sigma}(t)+\left(a(t)+B(t)-\dfrac{1}{R(t)}b^2(t)\pi(t)\right)\Sigma(t)+\dfrac{1}{R(t)}b(t)D(t)\pi(t)-A(t)\doteq 0,\\ \Sigma(T)=-F,\end{cases}\tag{5.29}$$

and then

$$\begin{cases} \dot{\theta}(t) + \left(a(t) - \dfrac{1}{R(t)}b^2(t)\pi(t)\right)\theta(t) - o(t)\Sigma(t) - \dfrac{1}{R(t)}b(t)r(t)\pi(t) \\ +\tilde{b}(t)\pi(t) + l(t) = 0, \\ \theta(T) = m. \end{cases} \tag{5.30}$$

It is clear that there are unique solutions to them, respectively. Thus we have

$$u(t) = \frac{1}{R(t)}[(D(t) - b(t)\Sigma(t))\hat{p}(t) - b(t)\pi(t)\hat{x}(t) - b(t)\theta(t) - r(t)].$$

The proof is then complete. □

Remark 5.1. The optimal control $u(\cdot)$ in Theorem 5.3 can be represented as the feedback of $(\hat{x}(\cdot), \hat{y}(\cdot))$ under some additional assumptions, say, $A(t) = D(t) = 0$.

Remark 5.2. The integrability condition in Definition 1.3 plays an important role in proving a density property of $\mathscr{U}_{ad}$. If $\mathscr{F}_t^{Y^v}$ does not depend on control or state, the integrability condition can be weakened. For example, let $f(t) = 0$ in Problem (FBLQ). It is easy to see that $\mathscr{U}_{ad} = \mathscr{U}_{ad}^0$ via

$$\mathscr{F}_t^{Y^v} = \mathscr{F}_t^{Y^0} = \sigma\{W(s); 0 \le s \le t\},$$

and thus, Lemma 5.3 holds automatically. Then the integrability condition can be relaxed to

$$\mathbb{E}\int_0^T v^2(s)ds < +\infty.$$

Moreover, all the results obtained in Sections 6.1.2–6.1.4 hold true under the weakened assumption.

5.2 An LQ Model of BSDE

This section is devoted to a special case of Problem (FBLQ), i.e., Problem (BLQ) introduced in Section 1.3. We adopt the assumptions and the notations introduced in Section 5.1 unless noted otherwise.

Problem (BLQ) supposes that $w(t)$ is observable at time t. It can be regarded as the case of (5.3) with $f(t) = g(t) = 0$ and $h(t) = 1$. Clearly,

$$\mathscr{F}_t^{Y^v} = \mathscr{F}_t^{Y^0} = \sigma\{W(s); 0 \le s \le t\},$$

i.e., $\mathscr{F}_t^{Y^v}$ is given a prior. Just as noted in Remark 5.2, the integrability condition in $\mathscr{U}_{ad}$ can be replaced by

$$\mathbb{E}\int_0^T v(t)^2 dt < +\infty.$$

Although the observation equation looks simple, the classical separation principle still does not hold, and hence, the resulting mathematical deductions are nontrivial in the fields of filtering and control. In order to obtain a unique closed-form solution, many theoretical results such as Theorems 5.1–5.3, Corollary 5.1, Lemmas 5.4, 5.5, and Remark 5.2 are used here. It should be emphasized that [31] first proposed the problem with $n = 0$. However, they were unable to completely solve the problem due to the limit of techniques used there. In detail, they were not able to include the important process $\tilde{C}(\cdot)\tilde{z}(\cdot)$ in the drift term of the state equation. The full case presented here is taken from [89].

We use four steps to solve the problem.

Step 1: Candidate optimal control.

According to Theorem 5.1, if $u(\cdot)$ is optimal, then it is necessary to satisfy

$$u(t) = \frac{1}{R(t)} D(t)\mathbb{E}\left[p(t)\middle|\mathscr{F}^Y(t)\right],$$

where $\mathscr{F}_t^Y = \sigma\{W(s); 0 \leq s \leq t\}$, and $(p(\cdot), y(\cdot), z(\cdot), \tilde{z}(\cdot)) \in \mathscr{L}^2_{\mathbf{F}^{W,\tilde{W}}}\left(0,T;\mathbb{R}^4\right)$ is the unique solution of the Hamiltonian system

$$\begin{cases} dp(t) = (B(t)p(t) - O(t)y(t))dt + C(t)p(t)dW(t) + \tilde{C}(t)p(t)d\tilde{W}(t), \\ -dy(t) = \left(B(t)y(t) + C(t)z(t) + \tilde{C}(t)\tilde{z}(t) + D(t)u(t)\right)dt \\ \qquad\qquad - z(t)dW(t) - \tilde{z}(t)d\tilde{W}(t), \\ p(0) = -Ny(0) - n, \quad y(T) = G. \end{cases} \tag{5.31}$$

Using Lemmas 5.4 and 5.5 to (5.31), we get the optimal filtering $(\hat{p}(t), \hat{y}(t), \hat{z}(t), \hat{\tilde{z}}(t))$ of $(p(t), y(t), z(t), \tilde{z}(t))$ with respect to $\mathscr{F}_t^Y$, which is governed by

$$\begin{cases} d\hat{p}(t) = (B(t)\hat{p}(t) - O(t)\hat{y}(t))dt + C(t)\hat{p}(t)dW(t), \\ -d\hat{y}(t) = \left(B(t)\hat{y}(t) + C(t)\hat{z}(t) + \tilde{C}(t)\hat{\tilde{z}}(t) + D(t)u(t)\right)dt - \hat{z}(t)dW(t), \\ \hat{p}(0) = -Ny(0) - n, \quad \hat{y}(T) = \hat{G}. \end{cases} \tag{5.32}$$

Note that, (5.32) is not a standard FBSDE because an additional filtering estimate $\hat{\tilde{z}}(\cdot)$ is contained. Therefore, its existence and uniqueness is not an immediate consequence of Lemma A.5.

Step 2: Existence and uniqueness of FBSDE (5.35).

We first introduce two differential equations

$$\begin{cases} \dot{\alpha}(t) - \left(2B(t) + C^2(t) + \tilde{C}^2(t)\right)\alpha(t) - \dfrac{1}{R(t)}D^2(t)\alpha^2(t) + O(t) = 0, \\ \alpha(0) = -N \end{cases} \tag{5.33}$$

and

$$\begin{cases} \dot{\beta}(t) - \left(B(t) + C^2(t) + \tilde{C}^2(t) + \dfrac{1}{R(t)}D^2(t)\alpha(t)\right)\beta(t) = 0, \\ \beta(0) = -n. \end{cases} \tag{5.34}$$

In fact, these two equations can be obtained similar to (5.28), (5.29), and (5.30). See also Step 3 of this example for more details.

It is well known that (5.33) and (5.34) admit unique solutions, which are uniformly bounded in $[0,T]$, respectively. Next, we introduce a standard FBSDE

$$\begin{cases} d\hat{p}(t) = (B(t)\hat{p}(t) - O(t)\hat{y}(t))dt + C(t)\hat{p}(t)dw(t), \\ -d\hat{y}(t) = \left[B(t)\hat{y}(t) + C(t)\hat{z}(t) + \left(\frac{1}{\alpha(t)}\tilde{C}^2(t) + \frac{1}{R(t)}D^2(t)\right)\hat{p}(t)\right]dt \\ \qquad - \hat{z}(t)dw(t), \\ \hat{p}(0) = -Ny(0) - n, \quad \hat{y}(T) = \hat{G}, \end{cases} \tag{5.35}$$

which is subject to an additional assumption condition as follows.

(H5.2) *The solution* $\alpha(\cdot)$ *of (5.33) satisfies*

$$\frac{1}{\alpha(t)}\tilde{C}^2(t) + \frac{1}{R(t)}D^2(t) \geq 0.$$

Clearly, (5.35) satisfies (H5.1). Thus, it follows from A.5 that (5.35) admits a unique solution $(\hat{p}(\cdot),\hat{y}(\cdot),\hat{z}(\cdot)) \in \mathscr{L}^2_{\mathbf{F}^Y}(0,T;\mathbb{R}^3)$.

Step 3: Equivalence between (5.32) and (5.35).

We first prove that the solution $(\hat{p}(\cdot),\hat{y}(\cdot),\hat{z}(\cdot))$ of (5.35) is a solution of (5.32). Set

$$u(t) = \frac{1}{R(t)}D(t)\hat{p}(t).$$

Then $u(\cdot)$ is an admissible control, and consequently, (5.31) admits a unique solution $(p(\cdot),y(\cdot),z(\cdot),\tilde{z}(\cdot)) \in \mathscr{L}^2_{\mathbf{F}^{W,\tilde{W}}}(0,T;\mathbb{R}^4)$. Similar to Theorem 5.3, set

$$p(t) = \alpha(t)y(t) + \beta(t) \text{ with } \alpha(0) = -N \text{ and } \beta(0) = -n, \tag{5.36}$$

where $\alpha(\cdot)$ and $\beta(\cdot)$ are deterministic and differentiable functions. Combining (5.31) with

$$u(t) = \frac{1}{R(t)}D(t)\hat{p}(t),$$

we get by Itô's formula

$$\begin{aligned} dp(t) &= \dot{\alpha}(t)y(t)dt + \alpha(t)dy(t) + \dot{\beta}(t)dt \\ &= \left[\dot{\alpha}(t)y(t) + \dot{\beta}(t)\right]dt \\ &\quad - \left(B(t)y(t) + C(t)z(t) + \tilde{C}(t)\tilde{z}(t) + \frac{1}{R(t)}D^2(t)\hat{p}(t)\right)\alpha(t)dt \\ &\quad + \alpha(t)z(t)dW(t) + \alpha(t)\tilde{z}(t)d\tilde{W}(t). \end{aligned}$$

Comparing the above equality with the SDE in (5.31), we have

$$\alpha(t)z(t) = C(t)p(t), \quad \alpha(t)\tilde{z}(t) = \tilde{C}(t)p(t) \tag{5.37}$$

and

$$\dot{\alpha}(t)y(t) - \left(B(t)y(t) + C(t)z(t) + \tilde{C}(t)\tilde{z}(t) + \frac{1}{R(t)}D^2(t)\hat{p}(t)\right)\alpha(t) + \dot{\beta}(t)$$
$$= B(t)p(t) - O(t)y(t). \tag{5.38}$$

Taking $\mathbb{E}\left[\cdot|\mathscr{F}^Y(t)\right]$ on both sides of (5.37) and (5.38), it yields

$$\alpha(t)\hat{z}(t) = C(t)\hat{p}(t), \quad \alpha(t)\hat{\tilde{z}}(t) = \tilde{C}(t)\hat{p}(t) \tag{5.39}$$

and

$$\dot{\alpha}(t)\hat{y}(t) - \left(B(t)\hat{y}(t) + C(t)\hat{z}(t) + \tilde{C}(t)\hat{\tilde{z}}(t) + \frac{1}{R(t)}D^2(t)\hat{p}(t)\right)\alpha(t) + \dot{\beta}(t)$$
$$= B(t)\hat{p}(t) - O(t)\hat{y}(t). \tag{5.40}$$

According to the second equation of (5.39), it is easy to see that the solution $(\hat{p}(\cdot), \hat{y}(\cdot), \hat{z}(\cdot))$ of (5.35) solves (5.32). By the way, (5.39) and (5.40) show us how to get (5.33) and (5.34).

Next, we prove that for fixed $u(\cdot)$, if (5.32) admits a solution in $\mathscr{L}^2_{\mathbf{F}^Y}\left(0, T; \mathbb{R}^4\right)$, then $(\hat{p}(\cdot), \hat{y}(\cdot), \hat{z}(\cdot), \hat{\tilde{z}}(\cdot))$ satisfies (5.35). Take

$$u(t) = \frac{1}{R(t)}D(t)\hat{p}(t).$$

Then there exists a unique solution $(p(\cdot), y(\cdot), z(\cdot), \tilde{z}(\cdot)) \in \mathscr{L}^2_{\mathscr{F}^{W,\tilde{W}}}\left(0, T; \mathbb{R}^4\right)$ to (5.31). Similar to the above analysis, it yields

$$\alpha(t)\hat{\tilde{z}}(t) = \tilde{C}(t)\hat{p}(t),$$

where $\alpha(\cdot)$ is the unique solution of (5.33). Putting

$$\hat{\tilde{z}}(t) = \frac{1}{\alpha(t)}\tilde{C}(t)\hat{p}(t) \text{ and } u(t) = \frac{1}{R(t)}D(t)\hat{p}(t)$$

into (5.32), we arrive at (5.35), which admits a unique solution $(\hat{p}(\cdot), \hat{y}(\cdot), \hat{z}(\cdot)) \in \mathscr{L}^2_{\mathbf{F}^Y}\left(0, T; \mathbb{R}^3\right)$. This shows that the solution $(\hat{p}(\cdot), \hat{y}(\cdot), \hat{z}(\cdot), \hat{\tilde{z}}(\cdot))$ of (5.32) is a solution of (5.35).

Therefore, the existence and uniqueness of (5.32) is equivalent to those of (5.35).

Step 4: Optimal feedback.

Define

$$u(t) = \frac{1}{R(t)}D(t)\mathbb{E}\left[p(t)|\mathscr{F}_t^Y\right] = \frac{1}{R(t)}D(t)\hat{p}(t), \tag{5.41}$$

where $(\hat{p}(\cdot), \hat{y}(\cdot), \hat{z}(\cdot))$ is the unique solution of (5.32). According to the existence of (5.32), (5.31) with (5.41) admits a unique solution $(p(\cdot), y(\cdot), z(\cdot))$. Theorem 5.2 and Corollary 5.1 imply that (5.41) is a unique optimal control. Inserting (5.36) into (5.41), it leads to

$$u(t) = \frac{1}{R(t)} D(t)(\alpha(t)\hat{y}(t) + \beta(t)), \tag{5.42}$$

where $\alpha(\cdot)$, $\beta(\cdot)$ and $(\hat{p}(\cdot), \hat{y}(\cdot), \hat{z}(\cdot))$ are the solutions of (5.33), (5.34) and (5.32).

We summarize the above analysis as follows.

Proposition 5.1. *Let (H5.2) hold. Then (5.41) is the unique optimal control of Problem (BLQ). Furthermore, (5.42) is its feedback representation.*

We emphasize that Problem (BLQ) is related to [31], in which the cost functional does not include $2ny^v(0)$; moreover, the uniqueness of optimal control was not proved.

5.3 An Optimal Premium Problem

The optimal premium problem introduced in Section 1.1, i.e., Problem (OP), is a special case of Problem (FBLQ), and hence, its optimal solution can be derived by Theorem 5.3. The main goal of this section is to provide a slightly different technique to solve Problem (OP) again. In detail, we first assume Problem (OP) is one with complete information for a while and seek the optimal solution under such assumption. Second, we construct a candidate optimal premium policy of the complete information problem and compute its optimal estimate by filtering theory of SDE. Last, we verify the filtering of the candidate optimal premium policy is indeed an optimal one of Problem (OP). This is the so-called *backward separation approach* introduced by Chapter 1.

Let $\hat{x}^v(t) = \mathbb{E}\left[x^v(t)|\mathscr{F}_t^{Y^v}\right]$ and $P(t) = \mathbb{E}\left[(x^v(t) - \hat{x}^v(t))^2|\mathscr{F}_t^{Y^v}\right]$ be the best estimate (in the sense of square error) and the mean squared error estimate. In terms of (1.4) and (1.5), we have

$$\begin{aligned} \hat{x}^v(t) &= \mathbb{E}\left[x_1^v(t) + x_2^v(t)|\mathscr{F}_t^{Y^v}\right] = \mathbb{E}\left[x_1^v(t)|\mathscr{F}_t^{Y_1}\right] + x_2^v(t), \\ P(t) &= \mathbb{E}\left[(x_1^v(t) - \hat{x}_1^v(t))^2|\mathscr{F}_t^{Y_1}\right]. \end{aligned}$$

Applying Theorem 2.1, we get

Corollary 5.2. *For any $v(\cdot) \in \mathscr{U}_{ad}$, the cash-balance process $x^v(\cdot)$ in (1.1) has a filtering estimate*

$$\begin{cases} d\hat{x}^v(t) = (\delta(t)\hat{x}^v(t) - b(t) + v(t))dt + (\sigma(t)\rho(t) + c\gamma(t)^{-1}P(t))d\hat{W}(t), \\ \hat{x}^v(0) = x_0, \end{cases} \tag{5.43}$$

where the stochastic process

$$\hat{W}(t) = \tilde{W}(t) + \int_0^t c\gamma^{-1}(s)(x^v(s) - \hat{x}^v(s))ds$$

is an observable 1-dimensional standard Brownian motion, and $P(\cdot)$ *satisfies*

$$\begin{cases} \dot{P}(t) - 2\delta(t)P(t) + (\sigma(t)\rho(t) + c\gamma^{-1}(t)P(t))^2 - \sigma^2(t) = 0, \\ P(0) = 0. \end{cases} \tag{5.44}$$

We shall now use the backward separation approach to solve Problem B. This can be illustrated by the following three steps.

Step 1: An optimal premium policy with complete information.

Suppose that Problem (OP) is one with complete information, for a while, in the sense that the premium policy $v(t)$ is $\mathscr{F}_t^W$-adapted. Using the maximum principle and Lagrangian method, we get the optimal premium policy with complete information

$$u(t) = -N^{-1}(t)e^{\beta t}(\phi(t)x(t) + \psi(t)),$$

where $x(\cdot)$, $\phi(\cdot)$ and $\psi(\cdot)$ satisfy

$$\begin{cases} dx(t) = \Big[(\delta(t) - N^{-1}(t)e^{\beta t}\phi(t))x(t) - b(t) - N^{-1}(t)e^{\beta t}\psi(t)\Big]\, dt \\ \qquad\qquad + \sigma(t)dW(t), \\ x(0) = x_0, \end{cases} \tag{5.45}$$

$$\begin{cases} \dot{\phi}(t) + 2\delta(t)\phi(t) - N^{-1}(t)e^{\beta t}\phi^2(t) + L(t)e^{-\beta t} = 0, \\ \phi(T) = Me^{-\beta T} \end{cases} \tag{5.46}$$

and

$$\begin{cases} \dot{\psi}(t) + (\delta(t) - N^{-1}(t)e^{\beta t}\phi(t))\psi(t) - b(t)\phi(t) - A(t)L(t)e^{-\beta t} = 0, \\ \psi(T) = \theta - c_0 Me^{-\beta T} \end{cases} \tag{5.47}$$

with

$$\theta = \frac{\int_0^T \Upsilon(t)dt + xe^{\int_0^T \varphi(t)dt} - c_0}{\int_0^T N^{-1}(t)e^{\beta t + 2\int_t^T \varphi(s)ds}dt},$$

$$\varphi(t) = \delta(t) - N^{-1}(t)e^{\beta t}\phi(t),$$

and

$$\Upsilon(t) = N^{-1}(t)\Big[\int_t^T \left(e^{\beta t}b(s)\phi(s) + A(s)L(s)e^{\beta(t-s)}\right)e^{\int_t^s \varphi(r)dr}ds \\ + c_0 Me^{\int_t^T(\varphi(s)-\beta)ds}\Big]e^{\int_t^T \varphi(s)ds} - b(t).$$

Step 2: A candidate optimal premium policy. Obviously, $\mathscr{U}_{ad}$ is a subset of the decision set of the counterpart of Problem (OP) with complete information. For Problem (OP), we cannot fully observe the cash-balance process $x^v(\cdot)$, but we can observe the stock price $S(\cdot)$ of the insurance firm. Our intuition is to replace the optimal cash-balance process $x(\cdot)$ by its filtering estimate $\hat{x}(\cdot)$ in the optimal premium policy of complete information. Based on this conjecture, we introduce a candidate optimal premium policy

$$\hat{u}(t) = -N^{-1}(t)e^{\beta t}(\phi(t)\hat{x}(t) + \psi(t)), \tag{5.48}$$

where $\phi(\cdot)$ and $\psi(\cdot)$ satisfy (5.46) and (5.47). According to (5.43), the filtering equation of $\hat{x}(\cdot)$ is

$$\begin{cases} d\hat{x}(t) = \Big[(\delta(t) - N^{-1}(t)e^{\beta t}\phi(t))\hat{x}(t) - b(t) - N^{-1}(t)e^{\beta t}\psi(t)\Big]\,dt \\ \qquad\qquad + \big(\sigma(t)\rho(t) + c\gamma^{-1}(t)p(t)\big)\,d\hat{W}(t), \\ \hat{x}(0) = x_0. \end{cases}$$

Step 3: Proof of optimization.

Since $\hat{x}^v(t) \perp (x^v(t) - \hat{x}^v(t))$, cost functional (1.6) can be rewritten as

$$\begin{aligned} J(v(\cdot)) &= \frac{1}{2}\mathbb{E}\Big\{\int_0^T e^{-\beta t}\left[L(t)(x^v(t) - \hat{x}^v(t) + \hat{x}^v(t) - A(t))^2 + N(t)v(t)^2\right]dt \\ &\quad + Me^{-\beta T}(x^v(T) - \hat{x}^v(T) + \hat{x}^v(T) - c_0)^2\Big\} \qquad (5.49) \\ &= \mathscr{J}(v(\cdot)) + \frac{1}{2}\left(\int_0^T e^{-\beta t}L(t)p(t)dt + Me^{-\beta T}p(t)\right) \end{aligned}$$

with

$$\begin{aligned} \mathscr{J}(v(\cdot)) &= \frac{1}{2}\mathbb{E}\Big\{\int_0^T e^{-\beta t}\left[L(t)(\hat{x}^v(t) - A(t))^2 + N(t)v^2(t)\right]dt \\ &\quad + Me^{-\beta T}(\hat{x}^v(T) - c_0)^2\Big\}, \end{aligned}$$

where $p(\cdot)$ is the solution of (5.44). Note that $p(\cdot)$ does not depend on the premium policy $v(\cdot)$. Then for any $v(\cdot) \in \mathscr{U}_{ad}$ it follows that

$$\begin{aligned} J(v(\cdot)) - J(\hat{u}(\cdot)) &= \frac{1}{2}\mathbb{E}\Big\{\int_0^T e^{-\beta t}\left[L(t)(\hat{x}^v(t) - \hat{x}(t))^2 + N(t)(v(t) - \hat{u}(t))^2\right]dt \\ &\quad + Me^{-\beta T}(\hat{x}^v(T) - \hat{x}(T))^2\Big\} + \Theta \end{aligned} \tag{5.50}$$

with

$$\begin{aligned} \Theta = \mathbb{E}\Big\{&\int_0^T e^{-\beta t}\left[L(t)(\hat{x}(t) - A(t))(\hat{x}^v(t) - \hat{x}(t)) + N(t)\hat{u}(t)(v(t) - \hat{u}(t))\right]dt \\ &+ Me^{-\beta T}(\hat{x}(T) - c_0)(\hat{x}^v(T) - \hat{x}(T))\Big\}. \end{aligned} \tag{5.51}$$

Since all the terms depending on $p(\cdot)$ have disappeared and the first term at the right-hand side of (5.50) is nonnegative, it is clear

$$J(v(\cdot)) - J(\hat{u}(\cdot)) \geq \Theta, \qquad \forall v(\cdot) \in \mathscr{U}_{ad}.$$

If $\Theta \equiv 0$ holds, then $\hat{u}(\cdot)$ defined by (5.48) is optimal. In fact, it follows from Itô's formula that

$$d(\phi(t)\hat{x}(t)+\psi(t)) = \Big[(A(t)-\hat{x}(t))L(t)e^{-\beta t} - \delta(t)(\phi(t)\hat{x}(t)+\psi(t))\Big]dt \\ +\phi(t)(\rho(t)\sigma(t)+c\gamma^{-1}(t)p(t))d\hat{W}(t)$$

and

$$\begin{aligned}
&d\left[(\phi(t)\hat{x}(t)+\psi(t))(\hat{x}^{v}(t)-\hat{x}(t))\right]\\
&=(\hat{x}^{v}(t)-\hat{x}(t))d(\phi(t)\hat{x}(t)+\psi(t))+(\phi(t)\hat{x}(t)+\psi(t))d(\hat{x}^{v}(t)-\hat{x}(t))\\
&=(\hat{x}^{v}(t)-\hat{x}(t))d(\phi(t)\hat{x}(t)+\psi(t))\\
&\quad+(\phi(t)\hat{x}(t)+\psi(t))\left[\delta(t)(\hat{x}^{v}(t)-\hat{x}(t))+v(t)-\hat{u}(t)\right]dt.
\end{aligned}$$

Note that $\mathbb{E}\hat{x}^{v}(T)=\mathbb{E}\hat{x}(T)=c_0$. Integrating from 0 to T and taking expectations on both sides of the foregoing equality,

$$\begin{aligned}
&\mathbb{E}\left[Me^{-\beta T}(\hat{x}(T)-c_0)(\hat{x}^{v}(T)-\hat{x}(T))\right]\\
&=\mathbb{E}\int_0^T(\hat{x}^{v}(t)-\hat{x}(t))d(\phi(t)\hat{x}(t)+\psi(t))\\
&\quad+\mathbb{E}\int_0^T(\phi(t)\hat{x}(t)+\psi(t))\left[\delta(t)(\hat{x}^{v}(t)-\hat{x}(t))+v(t)-\hat{u}(t)\right]dt.
\end{aligned}$$

Substituting the above two formulas into (5.51), we get $\Theta \equiv 0$.

Step 4: The optimal cost functional.

This step is similar to that of the complete information case. Thus we omit the detailed deductions and only give the key results.

Substituting (5.48) into (5.49), we derive

$$\begin{aligned}
J(\hat{u}(\cdot)) &= \frac{1}{2}\left(\int_0^T e^{-\beta t}L(t)p(t)dt+Me^{-\beta T}p(T)\right)\\
&\quad+\frac{1}{2}\left(\int_0^T e^{-\beta t}L(t)A^2(t)dt+Me^{-\beta T}c_0^2\right)\\
&\quad+\frac{1}{2}\mathbb{E}\Big\{\int_0^T e^{-\beta t}\Big[\left(L(t)+N^{-1}(t)e^{2\beta t}\phi^2(t)\right)\hat{x}^2(t)\\
&\quad-2\left(A(t)L(t)-N^{-1}(t)e^{2\beta t}\phi(t)\psi(t)\right)\hat{x}(t)\\
&\quad+N^{-1}(t)e^{2\beta t}\psi^2(t)\Big]dt+Me^{-\beta T}\hat{x}^2(T)-2c_0Me^{-\beta T}\hat{x}(T)\Big\}.
\end{aligned}$$

Applying Itô's formula to $\phi(\cdot)\hat{x}^2(\cdot)+2\psi(\cdot)\hat{x}(\cdot)$, we get

$$
\begin{aligned}
&\mathbb{E}\left(\phi(T)\hat{x}^2(T)+2\psi(T)\hat{x}(T)\right)\\
&=\int_0^T\left[\phi(t)\left(\rho(t)\sigma(t)+c\gamma^{-1}(t)p(t)\right)^2-2\psi(t)\left(b(t)+N^{-1}(t)e^{\beta t}\psi(t)\right)\right]dt\\
&\quad+2\mathbb{E}\int_0^T e^{-\beta t}\left(A(t)L(t)-N^{-1}(t)e^{2\beta t}\phi(t)\psi(t)\right)\hat{x}(t)dt\\
&\quad-\mathbb{E}\int_0^T e^{-\beta t}\left(L(t)+N^{-1}(t)e^{2\beta t}\phi^2(t)\right)\hat{x}^2(t)dt+\phi_0x_0^2+2\psi_0x_0.
\end{aligned}
$$

From the above equalities, then it follows

$$
\begin{aligned}
J(\hat{u}(\cdot))&=\frac{1}{2}\left(\int_0^T e^{-\beta t}L(t)A^2(t)dt+Me^{-\beta T}c_0^2\right)+\frac{1}{2}\phi_0x_0^2+\psi_0x_0-c_0\theta\\
&\quad+\frac{1}{2}\int_0^T\left[\phi(t)\left(\rho(t)\sigma(t)+c\gamma^{-1}(t)p(t)\right)^2\right]dt\\
&\quad+\frac{1}{2}\int_0^T\psi(t)\left(2b(t)-N^{-1}(t)e^{\beta t}\psi(t)\right)dt\\
&\quad+\frac{1}{2}\left(\int_0^T e^{-\beta t}L(t)p(t)dt+Me^{-\beta T}p(T)\right).
\end{aligned}
\tag{5.52}
$$

Now we conclude the aforementioned discussion with the following:

Theorem 5.4. *The optimal premium policy and the optimal cost functional of Problem (OP) are given explicitly by (5.48) and (5.52), respectively.*

If we remove the terminal constraint condition in Problem (OP), Problem (OP) is equivalent to minimizing (1.6) subject to (1.3), (1.8) and $\mathscr{U}_{ad}$. In this setting, we can obtain a result similar to Theorem 5.4 by setting $\theta=0$.

Note that the optimal premium policy with incomplete information has similar form to that with complete information, except that the actual state $x(\cdot)$ is replaced by its filter $\hat{x}(\cdot)$. Nevertheless, it doesn't imply that Problem (OP) is parallel to its counterpart with complete information. Instead, the study of Problem (OP) has its own intrinsic and unparalleled values from both economically illustrating and mathematically analyzing perspectives, as demonstrated as follows.

First, the optimal cost functional in incomplete information setup, given by (5.52), is quite different from that of complete information. This is mainly due to the essential change of observable filtration. Some interesting features raise when comparing these two functionals. For example, the quantity $p(\cdot)$ plays an important role in (5.52) whereas it is totally not involved in the complete information case. As to this feature, it is important to highlight that $p(\cdot)$ represents some quadratic deviation of filter to underlying state; thus it can be interpreted as some "information cost" to policy-makers due to their knowledge incompleteness. Intuitively, we can expect the "well-informed" policy-maker should outperform the "worsely informed" policy-maker because the latter is always subject to the "information cost" which will reduce the efficiency of optimal policy. In some extent, our study to Problem (OP) verifies this economic intuition from both qualitative and quantitative aspects by noting the terms involving $p(\cdot)$ in (5.52) are all nonnegative.

Second, besides its economic relevancy, Problem (OP) also has its mathematical significance. In fact, Problem (OP) is more difficult and challenging in theoretical analysis due to the lack of analytical tractability. Consequently, when handling the incomplete information in Problem (OP), we adopt a "decoupling technique," stochastic filtering and a backward separation technique which are not utilized in the study of the complete information case. These techniques can be viewed as by-products of Problem (OP) in the sense of mathematical analysis. These by-products, from our viewpoints, can be applied potentially to other interesting mathematical problems which are not limited in the present setup.

5.4 Notes

The earliest research on LQ problem can be traced back to Bellman et al. [4], Kalman [37], and Letov [44] for deterministic system; Kushner [41] and Wonham [96, 97] for Itô's stochastic system. There are a large volume of important works on this subject, see, e.g., Yong and Zhou [109], Tang [79], and literature cited therein for more information.

Different from the classical LQ control above, the LQ control of (fully coupled) FBSDE is almost an unexplored field. As far as we know, an early attempt was made by Dokuchaev and Zhou [17] and Wang et al. [83], where backward LQ control and forward-backward LQ control were considered, respectively. For further development, see, e.g., Lim and Zhou [48], Yong [107], Yu [110], Wang et al. [89], Li and Yu [47], Shi et al. [73, 74], and Wang et al. [92].

The results introduced in this chapter are taken mainly from Wang et al. [89]. The density argument in Problem (FBLQ) was inspired by Bensoussan [7]. See also Bensoussan and Viot [9] for early treatment. The decomposition technique used in Problem (FBLQ) and Problem (OP) plays an important role in decoupling the circular dependence of the control on the observation. The technique, however, is restricted to special structures of state and observation equations, say, the case that (5.1) and (5.3) are linear with respect to $(x^v(\cdot), y^v(\cdot), z^v(\cdot), \bar{z}^v(\cdot))$, the diffusion coefficient of (5.1) is deterministic, and the drift coefficient of (5.3) is independent of $(y^v(\cdot), z^v(\cdot), \bar{z}^v(\cdot))$. It is worth investigating the availability of the technique to decompose more general state and observation equations in the future.

Appendix A
BSDE and FBSDE

Nonlinear BSDE and fully coupled FBSDE have interesting applications in stochastic control, mathematical finance, and so on. Since the pioneering work of Pardoux and Peng [62], a large number of literature regarding BSDE and FBSDE have sprang up over the past two decades. The goal of this chapter is not to cover all the recent developments in BSDE and FBSDE, instead we only focus on the existence and uniqueness of the solution to BSDE and FBSDE.

A.1 BSDE

Consider a nonlinear BSDE

$$y(t) = \xi + \int_t^T g(s, y(s), z(s))ds - \int_t^T z(s)dW(s), \tag{A.1}$$

where $g : \Omega \times [0,T] \times \mathbf{R}^{n+n\times m} \to \mathbf{R}^n$ is a mapping; $W(\cdot)$ is an m-dimensional Brownian motion defined on the filtered probability space $(\Omega, \mathscr{F}, (\mathscr{F}_t)_{0\leq t\leq T}, \mathbf{P})$; $\mathscr{F}_t$ is the natural filtration generated by $W(t)$; and $\xi \in \mathscr{L}^2_{\mathscr{F}_T}(\Omega; \mathbf{R}^n)$. If there is a pair of processes $(y(\cdot), z(\cdot)) \in \mathscr{L}^2_{\mathrm{F}}(0,T; \mathbf{R}^{n+n\times m})$ which solves (A.1), we call $(y(\cdot), z(\cdot))$ an adapted solution of (A.1). Note that (A.1) is a kind of terminal value problem of SDE involving Itô's stochastic integral. However, we cannot simply reverse the time to get a solution for (A.1), because it would destroy the adaptiveness of the solution. This is essentially different from the case of ordinary differential equation (ODE).

To obtain the existence and uniqueness of the solution to (A.1), we make the assumptions below.

G. Wang et al., *An Introduction to Optimal Control of FBSDE with Incomplete Information*, SpringerBriefs in Mathematics,
https://doi.org/10.1007/978-3-319-79039-8

(Ha.1) For any $(y,z) \in \mathbf{R}^{n+n\times m}$, $g(\cdot,y,z) \in \mathscr{L}^2_{\mathbf{F}}(0,T;\mathbf{R}^n)$.

(Ha.2) There is a constant $C > 0$ such that for any $(\omega,t) \in \Omega \times [0,T]$, $(y_1,z_1) \in \mathbf{R}^{n+n\times m}$ and $(y_2,z_2) \in \mathbf{R}^{n+n\times m}$,

$$|g(t,y_1,z_1) - g(t,y_2,z_2)| \leq C(|y_1 - y_2| + |z_1 - z_2|).$$

Now we proceed to establishing a unique solution to BSDE (A.1). We will need a few lemmas.

Lemma A.1. *For any $y(0) \in \mathbf{R}^n$, $g \in \mathscr{L}^2_{\mathrm{F}}(0,T;\mathbf{R}^n)$ and $b \in \mathscr{L}^2_{\mathrm{F}}(0,T;\mathbf{R}^{n\times m})$, we define an Itô's process*

$$y(t) = y(0) + \int_0^t g(s)ds + \int_0^t b(s)dW(s).$$

Then, for any constant $\beta > 0$, we have

$$\begin{aligned}
&|y(0)|^2 + \mathbf{E}\int_0^T \left(\frac{\beta}{2}|y(t)|^2 + |b(t)|^2\right) e^{\beta t}dt \\
&\leq \mathbf{E}|y(T)|^2 e^{\beta T} + \frac{2}{\beta}\mathbf{E}\int_0^T |g(t)|^2 e^{\beta t}dt.
\end{aligned}$$

Proof. Applying Itô's formula to $|y(t)|^2 e^{\beta t}\rangle$, we derive

$$\begin{aligned}
e^{\beta T}|y(T)|^2 = |y(0)|^2 &+ \int_0^T |b(t)|^2 e^{\beta t}\rangle dt \\
&+ 2\int_0^T \langle g(t)dt + b(t)dW(t), y(t)e^{\beta t}\rangle \\
&+ \beta\int_0^T e^{\beta t}|y(t)|^2 dt.
\end{aligned}$$

Taking expectations on both sides of the above equation and noticing the elementary inequality $2ab \leq a^2 + b^2$, we arrive at

$$\begin{aligned}
&|y(0)|^2 + \mathbf{E}\int_0^T (\beta|y(t)|^2 + |b(t)|^2)\, e^{\beta t}dt \\
&= \mathbf{E}|y(T)|^2 e^{\beta T} - 2\mathbf{E}\int_0^T \langle g(t), y(t)e^{\beta t}\rangle dt \\
&\leq \mathbf{E}|y(T)|^2 e^{\beta T} + \mathbf{E}\int_0^T \left(\frac{2}{\beta}|g(t)|^2 + \frac{\beta}{2}|y(t)|^2\right) e^{\beta t}dt.
\end{aligned}$$

The proof is then completed. □

Lemma A.2. *For any $\xi \in \mathscr{L}^2_{\mathscr{F}_T}(\Omega;\mathbf{R}^n)$ and $g \in \mathscr{L}^2_{\mathrm{F}}(0,T;\mathbf{R}^n)$, there is a unique process $(y(\cdot),z(\cdot)) \in \mathscr{L}^2_{\mathrm{F}}(0,T;\mathbf{R}^{n+n\times m})$ which satisfies*

$$y(t) = \xi + \int_t^T g(s)ds - \int_t^T z(s)dW(s). \tag{A.2}$$

Proof. Equation (A.2) can be rewritten as

$$y(t) = y(0) - \int_0^t g(s)ds + \int_0^t z(s)dW(s).$$

First we prove the uniqueness of the solution. If there are two solutions, we denote the difference by $(\bar{y}(\cdot), \bar{z}(\cdot))$. Then,

$$\bar{y}(t) = \bar{y}(0) + \int_0^t \bar{z}(s)dW(s).$$

By Lemma A.1 with $\beta = 2$, we have

$$0 \leq |\bar{y}(0)|^2 + \mathbf{E}\int_0^T \left(|\bar{y}(t)|^2 + |\bar{z}(t)|^2\right) e^{2t}dt \leq 0.$$

This implies the uniqueness.

We now prove the existence. Let

$$y(t) = \mathbf{E}\left[\xi + \int_0^T g(s)ds|\mathscr{F}_t\right] - \int_0^t g(s)ds. \tag{A.3}$$

Since $\mathbf{E}\left[\xi + \int_0^T g(s)ds|\mathscr{F}_t\right]$ is a square-integrable martingale, there is a unique process $z(\cdot) \in \mathscr{L}^2_{\mathbf{F}}(0,T;\mathbf{R}^{n\times m})$ such that

$$y(t) = \mathbf{E}\left[\xi + \int_0^T g(s)ds\right] + \int_0^t z(s)dW(s) - \int_0^t g(s)ds. \tag{A.4}$$

Combining the equality (A.4) with the definition (A.3) of $y(\cdot)$, we have

$$\begin{aligned} y(T) &= \mathbf{E}\left[\xi + \int_0^T g(s)ds\right] + \int_0^T z(s)dW(s) - \int_0^T g(s)ds \\ &= \xi. \end{aligned} \tag{A.5}$$

Then (A.2) follows from (A.4) and (A.5), i.e., $(y(\cdot), z(\cdot)) \in \mathscr{L}^2_{\mathbf{F}}(0,T;\mathbf{R}^{n+n\times m})$ is a solution of (A.2). □

Theorem A.1. *Under (Ha.1)–(Ha.2), BSDE (A.1) admits a unique solution (y,z) in $\mathscr{L}^2_{\mathbf{F}}(0,T;\mathbf{R}^{n+n\times m})$.*

Proof. For any fixed $(y(\cdot), z(\cdot)) \in \mathscr{L}^2_{\mathbf{F}}(0,T;\mathbf{R}^{n+n\times m})$, it follows from (Ha.1) and (Ha.2) that $g(\cdot, y(\cdot), z(\cdot)) \in \mathscr{L}^2_{\mathbf{F}}(0,T;\mathbf{R}^n)$. Consider a linear BSDE

$$Y(t) = \xi + \int_t^T g(s, y(s), z(s))ds - \int_t^T Z(s)dW(s). \tag{A.6}$$

By Lemma A.2, (A.6) admits a unique solution $(Y(\cdot),Z(\cdot)) \in \mathscr{L}_{\mathrm{F}}^2(0,T;\mathbf{R}^{n+n\times m})$. Then we can define a mapping $I : \mathscr{L}_{\mathrm{F}}^2(0,T;\mathbf{R}^{n+n\times m}) \to \mathscr{L}_{\mathrm{F}}^2(0,T;\mathbf{R}^{n+n\times m})$ by $(y(\cdot),z(\cdot)) \mapsto (Y(\cdot),Z(\cdot))$ via (A.6). We are going to prove that for some $\beta > 0$, I is a contraction mapping on the metric space $\mathscr{L}_{\mathrm{F}}^2(0,T;\mathbf{R}^{n+n\times m})$ with metric ρ_β given by

$$\rho_\beta((Y_1,Z_1),(Y_2,Z_2))^2 = \mathbf{E}\int_0^T (|Y_1(t)-Y_2(t)|^2 + |Z_1(t)-Z_2(t)|^2)e^{\beta t}dt.$$

To this end, we take any $(y_1(\cdot),z_1(\cdot))$, $(y_2(\cdot),z_2(\cdot)) \in \mathscr{L}_{\mathrm{F}}^2(0,T;\mathbf{R}^{n+n\times m})$, and set

$$(Y_1(\cdot),Z_1(\cdot)) = I(y_1(\cdot),z_1(\cdot)), \quad (Y_2(\cdot),Z_2(\cdot)) = I(y_2(\cdot),z_2(\cdot)).$$

Let $\beta > 0$ be a constant to be decided later. Using (Ha.2) and Lemma A.1 to

$$\begin{aligned} Y_1(t) - Y_2(t) = &\int_t^T (g(s,y_1(s),z_1(s)) - g(s,y_2(s),z_2(s)))ds \\ &- \int_t^T (Z_1(s)-Z_2(s))dW(s), \end{aligned}$$

we arrive at

$$\begin{aligned} &\mathbf{E}\int_0^T \left(\frac{\beta}{2}|Y_1(t)-Y_2(t)|^2 + |Z_1(t)-Z_2(t)|^2\right)e^{\beta t}dt \\ &\le \frac{2}{\beta}\mathbf{E}\int_0^T |g(t,y_1(t),z_1(t)) - g(t,y_2(t),z_2(t))|^2 e^{\beta t}dt \\ &\le \frac{4C^2}{\beta}\mathbf{E}\int_0^T (|y_1(t)-y_2(t)|^2 + |z_1(t)-z_2(t)|^2)e^{\beta t}dt. \end{aligned}$$

Choose $\beta = \max\{2, 8C^2\}$. Then it is easy to see

$$\begin{aligned} &\mathbf{E}\int_0^T (|Y_1(t)-Y_2(t)|^2 + |Z_1(t)-Z_2(t)|^2)e^{\beta t}dt \\ &\le \frac{1}{2}\mathbf{E}\int_0^T (|y_1(t)-y_2(t)|^2 + |z_1(t)-z_2(t)|^2)e^{\beta t}dt. \end{aligned}$$

It implies the existence and uniqueness of solution to (A.6). □

A.2 FBSDE

Consider a fully coupled FBSDE

$$\begin{cases} dx(t) = b(t,x(t),y(t),z(t))dt + \sigma(t,x(t),y(t),z(t))dW(t), \\ -dy(t) = g(t,x(t),y(t),z(t))dt - z(t)dW(t), \\ x(0) = x_0, \quad y(T) = f(x(T)), \end{cases} \tag{A.7}$$

where $(b,\sigma,g):\ \Omega\times[0,T]\times\mathbf{R}^{n+m+m\times k}\to\mathbf{R}^{n+n\times k+m}$ and $f:\ \Omega\times\mathbf{R}^n\to\mathbf{R}^m$ are given mappings; $W(\cdot)$ is a k-dimensional standard Brownian motion defined on $(\Omega,\mathscr{F},(\mathscr{F}_t)_{0\le t\le T},\mathbf{P})$; $\mathscr{F}_t$ is the natural filtration generated by $W(t)$; and $x_0\in\mathrm{R}^n$. If there is a triple of processes $(x(\cdot),y(\cdot),z(\cdot))\in\mathscr{L}^2_{\mathscr{F}}(0,T;\mathbf{R}^{n+n\times m})$ which solves (A.7), we call $(x(\cdot),y(\cdot),z(\cdot))$ an adapted solution of (A.7).

For any $\lambda=(x,y,z)^\top$, we set $\Lambda(t,\lambda)=(-G^\top g,Gb,G\sigma)^\top(t,\lambda)$, where G is a given $m\times n$ full-rank matrix, and $G\sigma=(G\sigma_1,\cdots,G\sigma_k)$.

To establish the existence and uniqueness of the solution to the fully coupled FBSDE (A.7), we impose the following assumption.

(Ha.3) *For any $(t,\lambda)\in[0,T]\times\mathbf{R}^{n+n+n\times m}$, $\Lambda(t,\lambda)$ and $f(x)$ are uniformly Lipschitz with respect to λ and x, and $\Lambda(\cdot,\lambda)$ and $f(x)$ are in $\mathscr{L}^2_{\mathscr{F}}(0,T;\mathbf{R}^{n+n+n\times m})$ and $\mathscr{L}^2_{\mathrm{F}}(\Omega;\mathbf{R}^m)$, respectively.*

(Ha.4) *There are nonnegative constants μ_1, μ_2 and μ_3 such that for any $t\in[0,T]$, $\lambda,\bar\lambda\in\mathbf{R}^{n+n+n\times m}$,*

$$\langle\Lambda(t,\lambda)-\Lambda(t,\bar\lambda),\lambda-\bar\lambda\rangle\le-\mu_1|G(x-\bar x)|^2-\mu_2(|G^\top(y-\bar y)|^2+|G^\top(z-\bar z)|^2),$$
$$\langle f(x)-f(\bar x),G(x-\bar x)\rangle\ge\mu_3|G(x-\bar x)|^2,$$

with $\mu_1+\mu_2>0$ and $\mu_2+\mu_3>0$. Moreover, we have $\mu_1>0$, $\mu_3>0$ (resp., $\mu_2>0$) when $m>n$ (resp., $n>m$).

Remark A.1. If FBSDE (A.7) is decoupled, i.e., the coefficients b and σ do not depend on y and z, then the condition (Ha.3) is enough for the existence and uniqueness of the solution. However, for the fully coupled case, we need to impose the extra monotonicity condition (Ha.4).

Theorem A.2. *Under (Ha.3) and (Ha.4), FBSDE (A.7) has at most one solution in $\mathscr{L}^2_{\mathbf{F}}(0,T;\mathbf{R}^{n+m+m\times k})$.*

Proof. Assume that $\lambda(\cdot)=(x(\cdot),y(\cdot),z(\cdot))^\top$ and $\bar\lambda(\cdot)=(\bar x(\cdot),\bar y(\cdot),\bar z(\cdot))^\top$ are two solutions to (A.7). Set $\tilde\chi(\cdot)=\chi(\cdot)-\bar\chi(\cdot)$, where $\chi=\lambda,x,y,z$. Applying Itô's formula to $\langle G\tilde x(\cdot),\tilde y(\cdot)\rangle$, we get

$$\begin{aligned}&\mathbf{E}\langle f(x(T))-f(\bar x(T)),G\tilde x(T)\rangle-\mathbf{E}\langle\tilde y(t),G\tilde x(T)\rangle\\&=\mathbf{E}\int_t^T\langle\Lambda(s,\lambda(s))-\Lambda(s,\bar\lambda(s)),\tilde\lambda(s)\rangle ds\\&\le-\mu_1\mathbf{E}\int_t^T|G\tilde x(s)|^2ds-\mu_2\mathbf{E}\int_t^T|G^\top\tilde y(s)|^2ds.\end{aligned}$$

Noting that $\tilde x(0)=0$, it follows from the monotonicity condition (Ha.4) of f that

$$\mu_1\mathbf{E}\int_0^T|G\tilde x(t)|^2dt+\mu_2\mathbf{E}\int_0^T|G^\top\tilde y(t)|^2dt\le 0.$$

The rest of the proof is divided into three cases.

Case 1: $m>n$. In this case $\mu_1>0$, then $|G\tilde x(\cdot)|^2=0$. It implies that $x(\cdot)=\bar x(\cdot)$. In particular, $f(x(T))=f(\bar x(T))$, and hence, it follows from Theorem A.1 that $y(\cdot)=\bar y(\cdot)$ and $z(\cdot)=\bar z(\cdot)$.

Case 2: $m < n$. In this case $\mu_2 > 0$, and hence, $|G^T \tilde{y}(\cdot)|^2 = 0$. It implies that $y(\cdot) = \bar{y}(\cdot)$. Applying Itô's formula to $|\tilde{y}(\cdot)|^2 = 0$, we get

$$\int_0^T |z(t) - \bar{z}(t)|^2 dt = 0,$$

i.e., $z(\cdot) = \bar{z}(\cdot)$. Then it follows from the uniqueness of SDE that $x(\cdot) = \bar{x}(\cdot)$.

Case 3: $m = n$. Then, $\mu_1 > 0$ or $\mu_2 > 0$. The proofs follow from the same argument as that for Case 1 and Case 2, respectively. □

Next, we proceed to establishing the existence of a solution to FBSDE (A.7).

Lemma A.3. *Let κ be a nonnegative constant, $x \in \mathbf{R}^n$, $\phi \in \mathscr{L}_{\mathrm{F}}^2(0,T;\mathrm{R}^n)$, $\psi \in \mathscr{L}_{\mathrm{F}}^2(0,T;\mathbf{R}^{n\times k})$, $\gamma \in \mathscr{L}_{\mathrm{F}}^2(0,T;\mathbf{R}^m)$, and $\xi \in \mathscr{L}_{\mathscr{F}_T}^2(\Omega;\mathbb{R}^m)$. Then the linear FBSDE*

$$\begin{cases} dx(t) = \Big(-\mu_2 G^\top y(t) + \phi(t)\Big) dt + \Big(-\mu_2 G^\top z(t) + \psi(t)\Big) dW(t), \\ -dy(t) = (\mu_1 G x(t) + \gamma(t))dt - z(t)dW(t), \\ x(0) = x_0, \quad y(T) = \kappa G x(T) + \xi \end{cases}$$

has a unique solution $(x(\cdot), y(\cdot), z(\cdot)) \in \mathscr{L}_{\mathbf{F}}^2(0,T;\mathbf{R}^{n+m+m\times k})$.

Proof. First we assume $n \le m$. Since $G^\top G$ is strictly positive definite, we set

$$\begin{pmatrix} x_1 \\ y_1 \\ z_1 \end{pmatrix} = \begin{pmatrix} x \\ G^\top y \\ G^\top z \end{pmatrix}$$

and

$$\begin{pmatrix} y_2 \\ z_2 \end{pmatrix} = \begin{pmatrix} \big(I_m - G(G^\top G)^{-1} G^\top\big) y \\ \big(I_m - G(G^\top G)^{-1} G^\top\big) z \end{pmatrix}.$$

Multiplying $I_m - G(G^\top G)^{-1} G^\top$ on both sides of the BSDE, we get

$$\begin{cases} -dy_2(t) = \Big(I_m - G(G^\top G)^{-1} G^\top\Big) \gamma(t) dt - z_2(t) dW(t), \\ y_2(T) = \Big(I_m - G(G^\top G)^{-1} G^\top\Big) \xi, \end{cases}$$

where I_m denotes the $m \times m$ identity matrix. It follows from Theorem A.1 that $(y_2(\cdot), z_2(\cdot))$ is determined uniquely. Similarly, multiplying $G^\top$ on both sides of the same BSDE,

$$\begin{cases} dx_1(t) = (-\mu_2 y_1(t) + \phi(t))dt + (-\mu_2 z_1(t) + \psi(t))dW(t), \\ -dy_1(t) = \Big(\mu_1 G^\top G x_1(t) + G^\top \gamma(t)\Big) dt - z_1(t) dW(t), \\ x_1(0) = x_0, \quad y_1(T) = \kappa G^\top G x_1(T) + G^\top \xi. \end{cases} \tag{A.8}$$

To solve (A.8), we introduce a symmetric matrix-valued ODE

$$\begin{cases} \dot{K}(t) - \mu_2 K^2(t) + \mu_1 G^\top G = 0, \\ K(T) = \kappa G^\top G. \end{cases}$$

It is well known that the ODE has a unique solution. We then define a linear BSDE

$$\begin{cases} -dp(t) = \Big(-\mu_2 K(t)p(t) + K(t)\phi(t) + G^\top \gamma(t)\Big)\, dt \\ \qquad\qquad + (K(t)\psi(t) - (I_n + \mu_2 K(t))q(t))dW(t), \\ p(T) = G^\top \xi, \end{cases}$$

and an SDE

$$\begin{cases} dx_1(t) = [-\mu_2(K(t)x_1(t) + p(t)) + \phi(t)]dt + (\psi(t) - \mu_2(t)q(t))dW(t), \\ x_1(0) = x_0. \end{cases}$$

It is easy to check that

$$(x_1(\cdot), y_1(\cdot), z_1(\cdot)) = (x_1(\cdot), K(\cdot)x_1(\cdot) + p(\cdot), q(\cdot))$$

is the solution of (A.8). Once $(x_1(\cdot), y_1(\cdot), z_1(\cdot))$ and $(x_2(\cdot), y_2(\cdot))$ are resolved, then $(x(\cdot), y(\cdot), z(\cdot))$ is determined uniquely by

$$\begin{aligned} x(\cdot) &= x_1(\cdot), \\ y(\cdot) &= G(G^\top G)^{-1} y_1(\cdot) + y_2(\cdot), \\ z(\cdot) &= G(G^\top G)^{-1} z_1(\cdot) + z_2(\cdot). \end{aligned}$$

The proof for the case of $n > m$ is similar and we omit it. □

For any $\alpha \in [0,1]$, we introduce an auxiliary family of FBSDEs

$$\begin{cases} dx^\alpha(t) = \Big[(\alpha - 1)\mu_2 G^\top y^\alpha(t) + \alpha b(t, \lambda^\alpha(t)) + \phi(t)\Big]\, dt \\ \qquad\qquad + \Big[(\alpha - 1)\mu_2 G^\top z^\alpha(t) + \alpha\sigma(t, \lambda^\alpha(t)) + \psi(t)\Big]\, dW(t), \\ -dy^\alpha(t) = [(\alpha - 1)\mu_1 G x^\alpha(t) + \alpha g(t, \lambda^\alpha(t)) + \gamma(t)]dt - z^\alpha(t)dW(t), \\ x^\alpha(0) = x_0, \quad y^\alpha(T) = \alpha f(x^\alpha(T)) + (1-\alpha)Gx^\alpha(T) + \xi, \end{cases} \tag{A.9}$$

where $\phi,\ \psi,\ \gamma \in \mathscr{L}^2_{\mathrm{F}}(0,T;\mathbf{R}^{n+n\times k+m})$, and $\xi \in \mathscr{L}^2_{\mathscr{F}_T}(\Omega;\mathbf{R}^m)$. Note that (A.9) becomes (A.7) when $\alpha = 1$.

Lemma A.4. *Let (Ha.3) and (Ha.4) hold. Suppose that there exists $\alpha_0 \in [0,1)$ such that (A.9) with $\alpha = \alpha_0$ has a solution $(X^{\alpha_0}(\cdot), Y^{\alpha_0}(\cdot), Z^{\alpha_0}(\cdot))$. Then there exists a positive constant δ_0 (independent of α_0) such that for each $\delta \in [0,\delta_0]$, (A.9) with $\alpha = \alpha_0 + \delta$ has a solution $(X^{\alpha_0+\delta}(\cdot), Y^{\alpha_0+\delta}(\cdot), Z^{\alpha_0+\delta}(\cdot))$.*

Proof. Let δ_0 be a constant to be decided later, and let $\delta \in [0, \delta_0]$. For each $\lambda(\cdot) = (x(\cdot), y(\cdot), z(\cdot))^\top \in \mathscr{L}^2_{\mathrm{F}}(0,T;\mathbf{R}^{n+m+n\times m})$, we solve successively the FBSDE

$$\begin{cases} dX(t) = \Big[(\alpha_0 - 1)\mu_2 G^\top Y(t) + \alpha_0 b(t, U(t)) + \delta(\mu_2 G^\top y(t) + b(t,\lambda(t))) + \phi(t)\Big] dt \\ \qquad\qquad + \Big[(\alpha_0 - 1)\mu_2 G^\top Z(t) + \alpha_0 \sigma(t, U(t)) + \delta(\mu_2 G^\top z(t) + \sigma(t,\lambda(t))) + \psi(t)\Big] dW(t), \\ -dY(t) = [(\alpha_0 - 1)\mu_1 G X(t) + \alpha_0 g(t, U(t)) + \delta(g(t,\lambda(t)) - \mu_1 G x(t)) + \gamma(t)] dt - Z(t) dW(t), \\ X(0) = x_0, \quad Y(T) = \alpha_0 f(X(T)) + (1-\alpha_0) G X(T) + \delta(f(x(T)) - G x(T)) + \xi, \end{cases}$$

where $U(\cdot) = (X(\cdot), Y(\cdot), Z(\cdot))^\top$.

According to Theorem A.2 and the assumption of Lemma A.4, the above equation has a unique solution. We then define a mapping

$$\begin{aligned} &(U(\cdot),\, X(T)) = I_{\alpha_0+\delta}(\lambda(\cdot),\, x(T)) : \\ &\mathscr{L}^2_{\mathrm{F}}(0,T;\mathbf{R}^{n+m+m\times k}) \times \mathscr{L}^2_{\mathscr{F}_T}(\Omega;\mathbf{R}^n) \to \mathscr{L}^2_{\mathrm{F}}(0,T;\mathbf{R}^{n+m+m\times k}) \times \mathscr{L}^2_{\mathscr{F}_T}(\Omega;\mathbf{R}^n). \end{aligned}$$

We now prove that if δ is small enough, then the mapping is a contraction with respect to norm $\|\cdot\|$ given by

$$\|(U(\cdot),\, X(T))\|^2 = \mathbf{E}\left[\int_0^T |U(t)|^2 dt + |X(T)|^2\right].$$

Let $\bar\lambda(\cdot) = (\bar x(\cdot), \bar y(\cdot), \bar z(\cdot))^\top$ and $\bar U(\cdot) = (\bar X(\cdot), \bar Y(\cdot), \bar Z(\cdot))^\top$. Suppose that $\bar\lambda(\cdot) \in \mathscr{L}^2_{\mathrm{F}}(0,T;\mathbf{R}^{n+m+m\times k})$ and $\bar U(\cdot) \in \mathscr{L}^2_{\mathrm{F}}(0,T;\mathbf{R}^{n+m+m\times k})$. Set

$$\begin{aligned} \tilde\lambda(\cdot) &= (\tilde x(\cdot), \tilde y(\cdot), \tilde z(\cdot))^\top = (x(\cdot) - \bar x(\cdot), y(\cdot) - \bar y(\cdot), z(\cdot) - \bar z(\cdot))^\top, \\ \tilde U(\cdot) &= (\tilde X(\cdot), \tilde Y(\cdot), \tilde Z(\cdot))^\top = (X(\cdot) - \bar X(\cdot), Y(\cdot) - \bar Y(\cdot), Z(\cdot) - \bar Z(\cdot))^\top. \end{aligned}$$

Applying Itô's formula to $\langle G\tilde X(\cdot), \tilde Y(\cdot)\rangle$, we get

$$\begin{aligned} &\alpha_0 \mathbf{E}\langle f(X(T)) - f(\bar X(T)), G\tilde X(T)\rangle + (1-\alpha_0)\mathrm{E}\langle G\tilde X(T), G\tilde X(T)\rangle \\ &\quad + \delta \mathbf{E}\langle f(x(T)) - f(\bar x(T)) - G\tilde x(T), G\tilde x(T)\rangle \\ &= \mathbf{E}\int_0^T \langle \alpha_0(\Lambda(t, U(t)) - \Lambda(t, \bar U(t))), \tilde U(t)\rangle dt \\ &\quad - (1-\alpha_0)\mathbf{E}\int_0^T (\mu_1 \langle G\tilde X(t), G\tilde X(t)\rangle + \mu_2 \langle G^\top \tilde Y(t), G^\top \tilde Y(t)\rangle \\ &\quad + \mu_2 \langle G^\top \tilde Z(t), G^\top \tilde Z(t)\rangle) dt \\ &\quad + \delta \mathbf{E}\int_0^T (\mu_1 \langle G\tilde X(t), G\tilde x(t)\rangle + \mu_2 \langle G^\top \tilde Y(t), G^\top \tilde y(t)\rangle \\ &\quad + \mu_2 \langle G^\top \tilde Z(t), G^\top \tilde z(t)\rangle - \langle \tilde X(t), G^\top \tilde g(t)\rangle + \langle G^\top \tilde Y(t), \tilde b(t)\rangle + \langle \tilde Z(t), G\tilde\sigma(t)\rangle) dt, \end{aligned}$$

where $\tilde\chi(t) = \chi(t, \lambda(t)) - \chi(t, \bar\lambda(t))$ with $\chi = g, b, \sigma$. It follows from (Ha.3) and (Ha.4) that

$$
\begin{aligned}
&(\alpha_0\mu_3+1-\alpha_0)\mathbf{E}|G\tilde{X}(T)|^2+\mu_1\mathbf{E}\int_0^T|G\tilde{X}(t)|^2dt\\
&+\mu_2\mathbf{E}\int_0^T(|G^\top\tilde{Y}(t)|^2+|G^\top\tilde{Z}(t)|^2)dt\\
\leq\ &\delta K_1\mathbf{E}\left[\int_0^T(|\tilde{\lambda}(t)|^2+|U(t)|^2)dt+|\tilde{X}(T)|^2+|\tilde{x}(T)|^2\right].
\end{aligned}
$$

Similarly, applying usual techniques to $\langle\tilde{X}(\cdot),\tilde{X}(\cdot)\rangle$ and $\langle\tilde{Y}(\cdot),\tilde{Y}(\cdot)\rangle$, we derive the following estimates

$$
\sup_{0\leq t\leq T}\mathbf{E}|\tilde{X}(t)|^2\leq K_1\mathbf{E}\int_0^T(\delta|\tilde{\lambda}(t)|^2+|\tilde{Y}(t)|^2+|\tilde{Z}(t)|^2)dt,
$$

$$
\mathbf{E}\int_0^T|\tilde{X}(t)|^2dt\leq K_1T\mathbf{E}\int_0^T(\delta|\tilde{\lambda}(t)|^2+|\tilde{Y}(t)|^2+|\tilde{Z}(t)|^2)dt,
$$

$$
\begin{aligned}
\mathbf{E}\int_0^T(|\tilde{Y}(t)|^2+|\tilde{Z}(t)|^2)dt\leq\ &K_1\delta\mathbf{E}\left[\int_0^T|\tilde{\lambda}(t)|^2dt+|\tilde{x}(T)|^2\right]\\
&+K_1\mathbf{E}\left[\int_0^T|\tilde{X}(t)|^2dt+|\tilde{X}(T)|^2\right].
\end{aligned}
$$

Here the constant K_1 depends on the Lipschitz constants, G, μ_1, μ_2, and T. If $\mu_3>0$, then $\alpha_0\mu_3+1-\alpha_0\geq\mu$, $\mu=\min(1,\mu_3)>0$. Combining the above four estimates, then it is clear that, whatever $\mu_1>0$, $\mu_3>0$, $\mu_2\geq0$ or $\mu_1\geq0$, $\mu_3\geq0$, $\mu_2>0$, we always have

$$
\mathbf{E}\left[\int_0^T|\tilde{U}(t)|^2dt+|\tilde{X}(T)|^2\right]\leq K\delta\mathbf{E}\left[\int_0^T|\tilde{\lambda}(t)|^2dt+|\tilde{x}(T)|^2\right].
$$

Here the constant K depends on μ_1, μ_2, μ_3, K_1, and T. If we choose $\delta_0=\frac{1}{2K}$, then for any fixed $\delta\in[0,\delta_0]$, the mapping $I_{\alpha_0+\delta}$ is a contraction. It follows that this mapping has a unique fixed point $(X^{\alpha_0+\delta}(\cdot),Y^{\alpha_0+\delta}(\cdot),Z^{\alpha_0+\delta}(\cdot))$ which is the solution to (A.9) for $\alpha=\alpha_0+\delta$. Then the proof is completed. □

Theorem A.3. *Under (Ha.3) and (Ha.4), FBSDE (A.7) admits a unique solution in* $\mathscr{L}^2_{\mathrm{F}}(0,T;\mathbf{R}^{n+m+m\times k})$.

Proof. If we let $\kappa=1$ and $\xi=0$ in Lemma A.3, then it implies that (A.9) admits a unique solution when $\alpha=\alpha_0=0$. According to Lemma A.4, there exists a positive constant δ_0 depending on Lipschitz constants, μ_1, μ_2, μ_3, and T such that for each $\delta\in[0,\delta_0]$ there exists a unique solution to (A.9) with $\alpha=\alpha_0+\delta$. Repeat this process for N-times with $1\leq N\delta_0<1+\delta_0$. It follows that there exists a unique solution to (A.9) with $\alpha=1$ and $\xi=0$. Thus the proof is completed. □

The following FBSDE

$$\begin{cases} dx(t) = b(\omega,t,x(t),y(t),z(t))dt + \sigma(\omega,t,x(t),y(t),z(t))dW(t), \\ -dy(t) = g(\omega,t,x(t),y(t),z(t))dt - z(t)dW(t), \\ x(0) = \Psi(y(0)), \quad y(T) = \xi, \end{cases}$$

is slightly different from (A.7), where $(\omega,t,x,y,z) \in \Omega \times [0,T] \times \mathbb{R}^3$, $b : \Omega \times [0,T] \times \mathbb{R}^3 \to \mathbb{R}$, $\sigma : \Omega \times [0,T] \times \mathbb{R}^3 \to \mathbb{R}$, $g : \Omega \times [0,T] \times \mathbb{R}^3 \to \mathbb{R}$ and $\Psi : \mathbb{R} \to \mathbb{R}$ are four mappings. Define

$$\lambda = \begin{pmatrix} x \\ y \\ z \end{pmatrix} \text{ and } \Lambda(t,\lambda) = \begin{pmatrix} -g \\ b \\ \sigma \end{pmatrix} (\omega,t,\lambda).$$

To obtain a unique solution, we assume

(H5.1) *(i) $\Lambda(t,\lambda)$ and $\Psi(x)$ are uniformly Lipschitz with respect to λ and x, respectively. For each λ, $\Lambda(\cdot,\lambda)$ belongs to $\mathscr{L}^2_{\mathbf{F}^W}(0,T;\mathbb{R}^3)$ with*

$$\mathscr{F}^W(t) = \sigma\{W(s); 0 \le s \le t\}.$$

(ii)

$$\begin{cases} \langle \Lambda(t,\lambda) - \Lambda(t,\bar{\lambda}), \lambda - \bar{\lambda} \rangle \le -\mu_1 |x-\bar{x}|^2 - \mu_2(|y-\bar{y}|^2 + |z-\bar{z}|^2), \\ \langle \Psi(y) - \Psi(\bar{y}), y - \bar{y} \rangle \le -\mu_3 |y-\bar{y}|^2, \\ \forall \lambda = (x,y,z), \quad \bar{\lambda} = (\bar{x},\bar{y},\bar{z}), \end{cases}$$

where μ_1, μ_2, and μ_3 are nonnegative constants with $\mu_1 + \mu_2 > 0$ and $\mu_1 + \mu_3 > 0$.

Lemma A.5. *(Yu and Ji [107]) Let (H5.1) hold. The FBSDE admits a unique solution $(x(\cdot),y(\cdot),z(\cdot)) \in \mathscr{L}^2_{\mathbf{F}^W}(0,T;\mathbb{R}^3)$.*

A.3 Malliavin Derivatives

For the convenience of the reader, we collect some basic material about the Malliavin calculus needed in this book. Let $\Omega = C([0,T])$ and P the Wiener measure on Ω. Namely, $W(t) : \Omega \to \mathbf{R}$ given by $W(t,\omega) = \omega(t)$, $t \in [0,T]$ is a Brownian motion under P. Let $\mathscr{S}$ denote the set of random variables F of the form $F = \phi(W(h^1),\cdots,W(h^k))$, where $\phi \in C_b^\infty(\mathbf{R}^k)$, $h^i \in L^2([0,T])$, and $W(h^i) = \int_0^T h^i(s)dW(s)$. We define the Malliavin derivative of F as a stochastic process

$$D_t F = \sum_{j=1}^k \partial \phi_{x_j}(W(h^1),\cdots,W(h^k))h^j(t).$$

We define a norm $\|\cdot\|_{1,2}$ on $\mathscr{S}$ by

$$\|F\|_{1,2}^2 = \mathbf{E}\left[|F|^2 + \int_0^T |D_t F|^2 ds\right].$$

Let $\mathbf{D}_{1,2}$ be the completion of $\mathscr{S}$ with respect to the norm $\|\cdot\|_{1,2}$. It can be shown that D has a continuous extension to $\mathbf{D}_{1,2}$. Let $\mathbf{L}_{1,2}$ be the set of progressively measurable processes $\{F(t,\omega); 0 \le t \le T, \omega \in \Omega\}$ such that:

i) For $0 \le t \le T$, $F(t,\cdot) \in \mathbf{D}_{1,2}$;
ii) $D_t F \in \mathscr{L}^2_{\mathbf{F}}(0,T;\mathbf{R})$ admits a progressively measurable versions;
iii)

$$\|F\|_{1,2,a}^2 \equiv \mathrm{E}\left(\int_0^T |F(t,\omega)|^2 dt + \int_0^T \int_0^T |D_s F(t,\omega)|^2 ds dt\right) < \infty.$$

The following results can be found in Nualart [60]. We state them here without their proofs.

Lemma A.6. *If $F \in \mathrm{D}_{1,2}$ is $\mathscr{F}_s$-measurable, then $D_t F = 0$ for all $t > s$.*

Lemma A.7. *If $F \in \mathrm{D}_{1,2}$ and $\phi(\cdot) \in \mathscr{L}^2_{\mathrm{F}}(0,T;\mathbf{R})$, then*

$$\mathbf{E}\left(F\int_0^T \phi(t)dW(t)\right) = \mathbb{E}\int_0^T \phi(t)D_t F dt.$$

Lemma A.8. *If $F(t,\omega) \in \mathbb{L}_{1,2,}$, then*

$$\int_0^T F(t,\omega)dt,\ \int_0^T F(t,\omega)dW(t) \in \mathbb{D}_{1,2},$$

$$D_s \int_0^T F(t,\omega)dt = \int_\theta^T D_s F(s,\omega)ds,$$

$$D_s \int_0^T F(t,\omega)dW(t) = F(s,\omega) + \int_s^T D_s F(s,\omega)ds.$$

Finally, we consider the solution to the following BSDE

$$\begin{cases} -dY(t) = g(t,Y(t),Z(t))dt - Z(t)dW(t), \\ \quad Y(T) = F. \end{cases} \tag{A.10}$$

The following result can be found in El Karoui et al. [19].

Proposition A.1. *Suppose that $F \in \mathbf{D}_{1,2}$ and $g: \ \Omega \times [0,T] \times \mathrm{R}^2 \to \mathrm{R}$ is continuously differentiable in (y,z), with uniformly bounded and continuous derivatives and such that, for each (y,z), $g(\cdot,y,z)$ is in $\mathrm{L}_{1,2}$ with Malliavin derivative denoted by $D_s g(t,y,z)$. Let (Y,Z) be the solution of the BSDE (A.10). Also, suppose that*

- $$\mathbf{E}\left(\left|\int_0^T |g(t,0,0)|^2 dt\right|^2 + |F|^4\right) < \infty.$$

- $$\int_0^T \mathbf{E}(|D_sF|^2)ds + \int_0^T \mathbf{E}(|D_sg(t,Y,Z)|^2)ds < \infty,$$

  and for any $t \in [0,T]$ *and any* (y^1,z^1,y^2,z^2),

  $$|D_sg(t,y^1,z^1) - D_sg(t,y^2,z^2)| \le K_s(t)(|y^1-y^2| + |z^1-z^2|),$$

  where for a.e. $s \in [0,T]$, $\{K_s(t) : 0 \le t \le T\}$ *is an adapted process satisfying* $\mathbf{E}\int_0^T |K_s(t)|^2 dt < \infty$.

Then, $(Y,Z) \in \mathscr{L}^2_{\mathbf{F}}(0,T;\mathbf{D}^2_{1,2})$ *and*

$$D_sY(t) = D_sZ(t) = 0, \qquad 0 \le t < s \le T;$$

$$D_sY(t) = D_sF + \int_t^T \Big(g_y(r,Y(r),Z(r))D_sY(r) + g_z(r,Y(r),Z(r))D_sZ(r) \\ + D_sg(r,Y(r),Z(r))\Big)dr - \int_t^T D_sZ(r)dW(r), \qquad 0 \le t \le T.$$

Moreover, $\{D_tY(t) : 0 \le t \le T\}$ *is a version of* $\{Z(t) : 0 \le t \le T\}$.

A.4 Notes

The existence and uniqueness results provided here are mainly taken from Pardoux and Peng [62] and Peng and Wu [68]. Note that Condition (Ha.4) is only a sufficient one for the existence and uniqueness of solution to (A.7). For example, Theorem A.3 still holds true if (Ha.4) is replaced by the following condition.

(Ha.4)' *There are nonnegative constants* μ_1, μ_2, *and* μ_3 *such that for any* $t \in [0,T]$, $\lambda, \bar{\lambda} \in \mathbf{R}^{n+n+n\times m}$,

$$\langle \Lambda(t,\lambda) - \Lambda(t,\bar{\lambda}), \lambda - \bar{\lambda}\rangle \ge \mu_1|G(x-\bar{x})|^2 + \mu_2(|G^\top(y-\bar{y})|^2 + |G^\top(z-\bar{z})|^2), \\ \langle f(x) - f(\bar{x}), G(x-\bar{x})\rangle \le -\mu_3|G(x-\bar{x})|^2$$

with $\mu_1+\mu_2 > 0$ *and* $\mu_2+\mu_3 > 0$. *Moreover, we have* $\mu_1 > 0$, $\mu_3 > 0$ *(resp.,* $\mu_2 > 0$*) when* $m > n$ *(resp.,* $n > m$*).*

For other alternative sufficient conditions for Theorem A.3, we refer the reader to the recent work by Ma et al. [51]. About how to solve an FBSDE, please see Ma et al. [50] for the four-step scheme. See also, e.g., Cvitanić and Zhang [14], Zhao et al. [115], and literature cited therein for numerical approaches.

References

1. Arrow, K.J., Kurt, M.: Public Investment, the Rate of Return, and Optimal Fiscal Policy. John Hopkins Press, Baltimore (1970)
2. Bain, A., Crisan, D.: Fundamentals of Stochastic Filtering. Springer, New York (2009)
3. Bellman, R.: Dynamic Programming. Princeton University Press, Princeton (1957)
4. Bellman, R., Glicksberg, I., Gross, O.: Some Aspects of the Mathematical Theory of Control Processes. Rand Corporation, Santa Monica (1958)
5. Beneš, V.E.: Exact finite dimensional filters for certain diffusions with nonlinear drift. Stochastics **5**, 65–92 (1985)
6. Bensoussan, A.: Stochastic Control of Partially Observable Systems. Cambridge University Press, Cambridge (1992)
7. Bensoussan, A.: Stochastic Control by Functional Analysis Methods. North-Holland Publishing Company, New York (1982)
8. Bensoussan, A., Chutani, A., Sethi, S.P.: Optimal cash management under uncertainty. Oper. Res. Lett. **37**, 425–429 (2009)
9. Bensoussan, A., Viot, M.: Optimal control of stochastic linear distributed parameter systems. SIAM J. Control **13**, 904–926 (1975)
10. Bismut, J.M.: Analyse xonvexe et probabilités. Thèse. Faculté des Sciences de Paris, Paris (1973)
11. Bode, H.W., Shannon, C.E.: A simplified derivation of linear least square smoothing and prediction theory. Proc. IRE **38**, 417–425 (1950)
12. Boswijk, H., Hommes, C., Manzan, S.: Behavioral heterogeneity in stock prices. J. Econ. Dyn. Control **31**, 1938–1970 (2007)
13. Bouchard, B., Touzi, N., Zeghal, A.: Dual formulation of the utility maximization problem: the case of nonsmooth utility. Ann. Appl. Probab. **14**, 678–717 (2004)
14. Cvitanić, J., Zhang, J.: The steepest descent method for forward-backward SDEs. Electron. J. Probab. **10**, 1468–1495 (2005)
15. Davis, M.H.A.: The separation principle in stochastic control via Girsanov solutions. SIAM J. Control Optim. **14**, 176–188 (1976)
16. Davis, M.H.A., Varaiya, P.: Dynamic programming conditions for partially observable stochastic systems. SIAM J. Control **11**, 226–261 (1973)
17. Dokuchaev, N.G., Zhou, X.: Stochastic control problems with terminal contingent conditions. J. Math. Anal. Appl. **238**, 143–165 (1999)
18. Duncan, T.: Doctoral Dissertation. Dept. of Electrical Engineering, Stanford University (1967)

G. Wang et al., *An Introduction to Optimal Control of FBSDE with Incomplete Information*, SpringerBriefs in Mathematics,
https://doi.org/10.1007/978-3-319-79039-8

19. El Karoui, N., Peng, S., Quenez, M.C.: Backward stochastic differential equations in finance. Math. Finance **7**, 1–71 (1997)
20. Elliott, R.J., Aggoun, L., Moore, J.B.: Hidden Markov Models. Estimation and Control. Applications of Mathematics, vol. 29. Springer, New York (1995)
21. Fleming, W.H., Pardoux, E.: Optimal control for partially observed diffusions. SIAM J. Control Optim. **20**, 261–285 (1982)
22. Florentin, J.J.: Partial observability and optimal control. Int. J. Electron. **13**, 263–279 (1962)
23. Föllmer, H., Schied, A.: Convex measures of risk and trading constraints, Finance Stoch. **6**, 429–447 (2002)
24. Frittelli, M., Rosazza-Gianin, E.: Putting order in risk measures J. Bank. Finance **26**, 1473–1486 (2002)
25. Fujisaki, M., Kallianpur, G., Kunita, H.: Stochastic differential equations for the nonlinear filtering problem. Osaka J. Math. **9**, 19–40 (1972)
26. Hu, M.: Stochastic global maximum principle for optimization with recursive utilities. Probab. Uncertain. Quant. Risk **2**, 1–20 (2017)
27. Hu, Y., Øksendal, B.: Partial information linear quadratic control for jump diffusions. SIAM J. Control Optim. **47**, 1744–1761 (2008)
28. Hu, Y., Peng, S.: Solution of forward-backward stochastic differential equations. Probab. Theory Rel. Fields **103**, 273–283 (1995)
29. Huang, J., Shi, J.: Maximum principle for optimal control of fully coupled forward-backward stochastic differential delayed equations. ESAIM Control Optim. Calc. Var. **18**, 1073–1096 (2012)
30. Huang, J., Wang, G., Wu, Z.: Optimal premium policy of an insurance firm: full and partial information. Insur. Math. Econ. **47**, 208–215 (2010)
31. Huang, J., Wang, G., Xiong, J.: A maximum principle for partial information backward stochastic control problems with applications. SIAM J. Control Optim. **48**, 2106–2117 (2009)
32. Ji, S., Wei, Q.: A maximum principle for fully coupled forward-backward stochastic control systems with terminal state constraints. J. Math. Anal. Appl. **407**, 200–210 (2013)
33. Joseph, D.P., Tou, T.J.: On linear control theory. Trans. Am. Inst. Electr. Eng. Part II Appl. Ind. **80**, 193–196 (1961)
34. Kailath, T.: An innovation approach to least-square estimation. Part I: Linear filtering in additive white noise. IEEE Trans. Automat. Control **13**, 646–655 (1968)
35. Kailath, T., Frost, P.: An innovation approach to least-square estimation. Part II: Linear smoothing in additive white noise. IEEE Trans. Automat. Control **13**, 656–660 (1968)
36. Kallianpur, G.: Stochastic Filtering Theory. Springer, New York (1980)
37. Kalman, R.E.: Contributions to the theory of optimal control. Bol. Soc. Mat. Mexicana **5**, 102–119 (1960)
38. Kalman, R.E., Bucy, R.S.: New results in linear filtering and prediction theory. Trans. ASME Ser. D (J. Basic Eng. ASME) **83**, 95–108 (1961)
39. Kalman, R.E., Koepcke, R.W.: Optimal synthesis of linear sampling control systems using generalized performance indexes. Trans. ASME **80**, 1820–1826 (1958)
40. Kushner, H.J.: On the differential equations satisfied by conditional probablitity densities of Markov processes with applications. SIAM J. Control **2**, 106–119 (1962)
41. Kushner, H.J.: Optimal stochastic control. IRE Trans. Automat. Control **7**, 120–122 (1962)
42. Kusher, H.J., Schweppe, F.C.: A maximum principle for stochastic control systems. J. Math. Anal. Appl. **8**, 287–302 (1964)
43. Lakner, P.: Optimal trading strategy for an investor: the case of partial information. Stoch. Proc. Appl. **76**, 77–97 (1998)
44. Letov, A.M.: Analytic design of regulators. Avtomat. Telemekh. 436–446, 561–571, 661–669 (1960, in Russian); English Transl.: Automat. Remote Control 21 (1960)
45. Li, X., Tang, S.: General necessary conditions for partially observed optimal stochastic controls. J. Appl. Prob. **32**, 1118–1137 (1995)

46. Li, J., Wei, Q.: Optimal control problems of fully coupled FBSDEs and viscosity solutions of Hamilton-Jacobi-Bellman equations. SIAM J. Control Optim. **52**, 1622–1662 (2014)
47. Li, N., Yu, Z.: Recursive stochastic linear-quadratic optimal control and nonzero-sum differential game problems with random jumps. Adv. Differ. Equ. **144** (2015). https://doi.org/10.1186/s13662-015-0439-1
48. Lim, A.E.B., Zhou, X.: Linear-quadratic control of backward stochastic differential equations. SIAM J. Control Optim. **40**, 450–474 (2001)
49. Liptser, R.S., Shiryayev, A.N.: Statistics of Random Processes. Springer, New York (1977)
50. Ma, J., Protter, P., Yong, J.: Solving forward-backward stochastic differential equations explicitly—a four step scheme. Probab. Theory Rel. Fields **98**, 339–359 (1994)
51. Ma, J., Wu, Z., Zhang, D., Zhang, J.: On well-posedness of forward-backward SDEs–a unified approach. Ann. Appl. Probab. **25**, 2168–2214 (2015)
52. Ma, J., Yong, J.: Forward-Backward Stochastic Differential Equations and Their Applications. Lecture Notes in Mathematics, vol. 1702. Springer, New York (1999)
53. Mangasarian, O.L.: Sufficient conditions for the optimal control of nonlinear systems. SIAM J. Control Optim. **32**, 139–152 (1966)
54. Meng, Q.: Maximum principle for optimal control problem of fully coupled forward-backward stochastic systems with partial information. Sci. China Ser. A Math. **52**, 1579–1588 (2009)
55. Menaldi, J.L.: The separation principle for impulse control problems. Proc. Am. Math. Soc. **82**, 439–445 (1981)
56. Mortensen, R.E.: Doctoral Dissertation. Dept. of Electrical Engineering. University of California at Berkeley (1966)
57. Nagai, H., Peng, S.: Risk-sensitive dynamic portfolio optimization with partial information on infinite time horizon. Ann. Appl. Probab. **12**, 173–195 (2002)
58. Nie, T., Shi, J., Wu, Z.: Connection between MP and DPP for stochastic recursive optimal control problems: viscosity solution framework in the general case. SIAM J. Control Optim. **55**, 3258–3294 (2017)
59. Norberg, R.: Ruin problems with assets and liabilities of diffusion type. Stoch. Proc. Appl. **81**, 255–269 (1999)
60. Nualart, D.: The Malliavin Calculus and Related Topics. Springer, Berlin (1995)
61. Økesandal, B., Sulem, A.: Maximum principle for optimal control of stochastic differential equations with applications. SIAM J. Control Optim. **48**, 2945–2976 (2010)
62. Pardoux, E., Peng, S.: Adapted solution of a backward stochastic differential equation. Syst. Control Lett. **44**, 55–61 (1990)
63. Peng, S.: A generalized dynamic programming principle and Hamilton-Jacobi-Bellmen equation. Stoch. Stoch. Rep. **38**, 119–134 (1992)
64. Peng, S.: Backward stochastic differential equations and applications to optimal control. Appl. Math. Optim. **27**, 125–144 (1993)
65. Peng, S.: Backward stochastic differential equations-stochastic optimization theory and viscosity solutions of HJB equations. In: Yan, J., Peng, S., Fang, S., Wu, L. (eds.) Topics on Stochastic Analysis, pp. 85–138. Science Press, Beijing (1997, in Chinese)
66. Peng, S.: A general stochastic maximum principle for optimal control problems. SIAM J. Control Optim. **28**, 966–979 (1990)
67. Peng, S.: Backward stochastic differential equations and applications to optimal control. Appl. Math. Optim. **27**, 125–144 (1993)
68. Peng, S., Wu, Z.: Fully coupled forward-backward stochastic differential equations and applications to optimal control. SIAM J. Control Optim. **37**, 825–843 (1999)
69. Pham, H.: Continuous-Time Stochastic Control and Optimization with Financial Applications. Stochastic Modelling and Applied Probability, vol. 61. Springer, Berlin (2009)
70. Pontryagin, L.S., Boltyanski, V.G., Gamkrelidze, R.V., Mischenko, E.F.: Mathematical Theory of Optimal Processes. Wiley, New York (1962)
71. Rosazza-Gianin, E.: Risk measures via g-expectations. Insur. Math. Econ. **39**, 19–34 (2006)

72. Shen, B., Wang, Z., Shu, H.: Nonlinear Stochastic Systems with Incomplete Information. Filtering and Control. Springer, London (2013)
73. Shi, J., Wang, G., Xiong, J.: Leader-follower stochastic differential game with asymmetric information and applications. Automat. J. IFAC **63**, 60–73 (2016)
74. Shi, J., Wang, G., Xiong, J.: Linear-quadratic stochastic Stackelberg differential game with asymmetric information. Sci. China Inform. Sci. **60**, 092202:1–092202:15 (2017)
75. Shi, J., Wu, Z.: The maximum principle for fully coupled forward-backward stochastic control system. Acta Automat. Sinica **32**, 161–169 (2006)
76. Simon, H.A.: Dynamic programming under uncertainty with a quadratic criterion function. Econometrica **24**, 74–81 (1956)
77. Stratonovich, R.L.: Conditional Markov processes. Theory Prob. Appl. **5**, 156–178 (1960)
78. Tang, S.: The maximum principle for partially observed optimal control of stochastic differential equations. SIAM J. Control Optim. **36**, 1596–1617 (1998)
79. Tang, S.: General linear quadratic optimal stochastic control problems with random coefficients: linear stochastic Hamilton systems and backward stochastic Riccati equations. SIAM J. Control Optim. **42**, 53–75 (2003)
80. Touzi, N.: Stochastic Control Problems, Viscosity Solutions and Application to Finance. Scuola Normale Superiore di Pisa, Quaderni (2004)
81. Touzi, N.: Optimal Stochastic Control, Stochastic Target Problems, and Backward SDE. With Chapter 13 by Angès Tourin. Fields Institute Monographs, vol. 29. Springer, New York; Fields Institute for Research in Mathematical Sciences, Toronto (2013)
82. Wang, G.: Partially Observed Stochastic Control Systems and Their Applications. Doctoral Dissertation. School of Mathematics and System Science, Shandong University (2007, in Chinese)
83. Wang, X., Gao, Z., Wu, Z.: Forward-backward stochastic differential equation and the linear quadratic stochastic optimal control. Acta Automat. Sinica **29**, 32–37 (2003)
84. Wang, G., Wu, Z.: Kalman-Bucy filtering equations of forward and backward stochastic systems and applications to recursive optimal control problems. J. Math. Anal. Appl. **342**, 1280–1296 (2008)
85. Wang, G., Wu, Z.: The maximum principles for stochastic recursive optimal control problems under partial information. IEEE Trans. Automat. Control **54**, 1230–1242 (2009)
86. Wang, G., Wu, Z.: General maximum principles for partially observed risk-sensitive optimal control problems and applications to finance. J. Optim. Theory Appl. **141**, 677–700 (2009)
87. Wang, G., Wu, Z.: Mean-variance hedging and forward-backward stochastic differential filtering equations. Abstr. Appl. Anal. **2011**, Art. ID 310910, 20 pp. (2011)
88. Wang, G., Wu, Z., Xiong, J.: Maximum principles for forward-backward stochastic control systems with correlated state and observation noises. SIAM J. Control Optim. **51**, 491–524 (2013)
89. Wang, G., Wu, Z., Xiong, J.: A linear-quadratic optimal control problem of forward-backward stochastic differential equations with partial information. IEEE Trans. Automat. Control **60**, 2904–2916 (2015)
90. Wang, G., Xiao, H.: Arrow sufficient conditions for optimality of fully coupled forward-backward stochastic differential equations with applications to finance. J. Optim. Theory Appl. **165**, 639–656 (2015)
91. Wang, G., Xiao, H., Xing, G.: An optimal control problem for mean-field forward-backward stochastic differential equation with noisy. Automatica J. IFAC **86**, 104–109 (2017)
92. Wang, G., Xiao, H., Xiong, J.: Linear-quadratic non-zero sum differential games of backward stochastic differential equations with asymmetric information. http://arxiv.org/abs/1407.0430 (2016). Accessed 9 Mar 2016
93. Wang, G., Xiong, J., Zhang, S.: Partially observable stochastic optimal control. Int. J. Numer. Anal. Model. **13**, 493–512 (2016)
94. Wang, G., Yu, Z.: A partial information non-zero sum differential game of backward stochastic differential equations with applications. Automat. J. IFAC **48**, 342–352 (2012)

95. Wang, G., Zhang, C., Zhang, W.: Stochastic maximum principle for mean-field type optimal control with partial information. IEEE Trans. Automat. Control **59**, 522–528 (2014)
96. Wonham, W.M.: On the separation theorem of stochastic control. SIAM J. Control **6**, 312–326 (1968)
97. Wonham, W.M.: On a matrix Riccati equation of stochastic control. SIAM J. Control 6, 681–697 (1968)
98. Wonham, W.M.: Some applications of stochastic differential equations to optimal nonlinear filtering. SIAM J. Control **2**, 347–369 (1965) (1965)
99. Wu, Z.: Maximum principle for optimal control problem of fully coupled forward-backward stochastic systems. Syst. Sci. Math. Sci. **11**, 249–259 (1998)
100. Wu, Z.: A maximum principle for partially observed optimal control of forward-backward stochastic control systems. Sci. China Ser. F Inf. Sci. **53**, 1–10 (2010)
101. Wu, Z.: A general maximum principle for optimal control of forward-backward stochastic systems. Automat. J. IFAC **49**, 1473–1480 (2013)
102. Wu, Z., Yu, Z.: Dynamic programming principle for one kind of stochastic recursive optimal control problem and Hamilton-Jacobi-Bellman equation. SIAM J. Control Optim. **47**, 2616–2641 (2008)
103. Xiao, H.: The maximum principle for partially observed optimal control of forward-backward stochastic systems with random jumps. J. Syst. Sci. Complex **24**, 1083–1099 (2011)
104. Xiong, J.: An Introduction to Stochastic Filtering Theory. Oxford University Press, London (2008)
105. Xiong, J., Zhou, X.: Mean-variance portfolio selection under partial information. SIAM J. Control Optim. **46**, 156–175 (2007)
106. Xu, W.: Stochastic maximum principle for optimal control problem of forward and backward system. J. Aust. Math. Soc. Ser. B **37**, 172–185 (1995)
107. Yong, J.: Optimality variational principle for controlled forward-backward stochastic differential equations with mixed initial-terminal conditions. SIAM J. Control Optim. **48**, 4119–4156 (2010)
108. Yong, J.: A leader-follower stochastic linear quadratic differential game. SIAM J. Control Optim. **41**, 1015–1041 (2002)
109. Yong, J., Zhou, X.: Stochastic Controls: Hamiltonian Systems and HJB Equations. Springer, New York (1999)
110. Yu, Z.: Linear-quadratic optimal control and nonzero-sum differential game of forward-backward stochastic system. Asian J. Control **14**, 173–185 (2012)
111. Yu, Z., Ji, S.: Linear-quadratic nonzero-sum differential game of backward stochastic differential equations. In: Proceedings of the 27th Chinese Control Conference, July 16–18, 2008, Kunming, Yunnan, pp. 562–566 (2008)
112. Zakai, M.: On the optimal filtering of diffusion processes. Z. Wahrsch. Geb. **11**, 230–243 (1969)
113. Zhang, H., Xie, L.: Control and estimation of systems with input/output delays. Lecture Notes in Control and Information Sciences, vol. 355. Springer, Berlin (2007)
114. Zhang, Q.: Controlled partially observed diffusions with correlated noise. Appl. Math. Optim. **22**, 265–285 (1990)
115. Zhao, W., Fu, Y., Zhou, T.: New kinds of high-order multistep schemes for coupled forward backward stochastic differential equations. SIAM J. Sci. Comput. **36**, A1731–A1751 (2014)
116. Zhou, X.Y.: Sufficient conditions of optimality for stochastic systems with controllable diffusions. IEEE Trans. Automat. Control **40**, 1176–1179 (1996)

Index

G. Wang et al., *An Introduction to Optimal Control of FBSDE with Incomplete Information*, SpringerBriefs in Mathematics,
https://doi.org/10.1007/978-3-319-79039-8

Printed in the United States
By Bookmasters